MODFLOW-2005, THE U.S. GEOLOGICAL SURVEY MODULAR GROUND-WATER MODEL – DOCUMENTATION OF SHARED NODE LOCAL GRID REFINEMENT (LGR) AND THE BOUNDARY FLOW AND HEAD (BFH) PACKAGE

By Steffen W. Mehl and Mary C. Hill

Chapter 12 of
Book 6, Modeling Techniques
Section A, Ground Water

Prepared in cooperation with the

U.S. Geological Survey Office of Ground Water and U.S. Department of Energy

Techniques and Methods 6-A12

U.S. Department of the Interior
U.S. Geological Survey

U.S. Department of the Interior
Gale A. Norton, Secretary

U.S. Geological Survey
P. Patrick Leahy, Acting Director

U.S. Geological Survey, Denver, Colorado 2005

For sale by U.S. Geological Survey, Information Services
Box 25286, Denver Federal Center
Denver, CO 80225

For more information about the USGS and its products:
Telephone: 1-888-ASK-USGS
World Wide Web: http://www.usgs.gov/

Preface

This report describes shared node Local Grid Refinement (LGR) for MODFLOW-2005, the U.S. Geological Survey's three-dimensional finite-difference ground-water model. LGR is designed to allow users to create MODFLOW simulations using a refined grid that is embedded within a coarser grid.

This report also describes the Boundary Flow and Head (BFH) Package for MODFLOW-2005. The BFH Package allows the refined grid to be run separately from the rest of the model, and the rest of the model to be run separately from the refined grid.

The performance of the programs has been tested in a variety of applications. Future applications, however, might reveal errors that were not detected in the test simulations. Users are requested to notify the U.S. Geological Survey of any errors found in this document or the computer program using the e-mail address available at the web address below. Updates might occasionally be made to both this document and to LGR and the BFH Package. Users can check for updates on the Internet at URL http://water.usgs.gov/software/ground_water.html/.

Contents

FIGURES

TABLES

Conversion Factors

Multiply	By	To obtain
meter (m)	3.281	foot
meter per day (m/day)	3.281	foot per day
square meter per second (m^2/s)	10.76	square foot per day
square meter per hour (m^2/hr)	10.76	square foot per hour
cubic meter per second (m^3/s)	35.31	cubic foot per second
cubic meter per hour (m^3/hr)	35.31	cubic foot per hour

Acronyms

MODFLOW-2005 Packages and capabilities

BCF	Block-Centered Flow
BFH	Boundary Flow and Head
CHD	Constant-Head Boundary
DE4	Direct solution based on alternating diagonal ordering
DIS	Discretization
DRN	Drain
EVT	Evapotranspiration
FHB	Flow and Head Boundary
GHB	General-Head Boundary
GMG	Geometric Multigrid
HFB	Horizontal-Flow Barrier
HUF	Hydrologic-Unit Flow
LMG	Link Algebraic Multigrid
LPF	Layer Property Flow
LVDA	Layer Variable-Direction Horizontal Anisotropy
MNW	Multi-Node Well
PCG	Preconditioned Conjugate-Gradient
RCH	Recharge
RIV	River
SIP	Strongly Implicit Procedure
STR	Stream
WEL	Well

Other acronyms

LGR	Local Grid Refinement
TMR	Telescopic Mesh Refinement
SEN	Sensitivity Process
PES	Parameter-Estimation Process

MODFLOW-2005, THE U.S. GEOLOGICAL SURVEY MODULAR GROUND-WATER MODEL – DOCUMENTATION OF SHARED NODE LOCAL GRID REFINEMENT (LGR) AND THE BOUNDARY FLOW AND HEAD (BFH) PACKAGE

By Steffen W. Mehl and Mary C. Hill

Abstract

This report documents the addition of shared node Local Grid Refinement (LGR) to MODFLOW-2005, the U.S. Geological Survey modular, transient, three-dimensional, finite-difference ground-water flow model. LGR provides the capability to simulate ground-water flow using one block-shaped higher-resolution local grid (a child model) within a coarser-grid parent model. LGR accomplishes this by iteratively coupling two separate MODFLOW-2005 models such that heads and fluxes are balanced across the shared interfacing boundary. LGR can be used in two-and three-dimensional, steady-state and transient simulations and for simulations of confined and unconfined ground-water systems.

Traditional one-way coupled telescopic mesh refinement (TMR) methods can have large, often undetected, inconsistencies in heads and fluxes across the interface between two model grids. The iteratively coupled shared-node method of LGR provides a more rigorous coupling in which the solution accuracy is controlled by convergence criteria defined by the user. In realistic problems, this can result in substantially more accurate solutions and require an increase in computer processing time. The rigorous coupling enables sensitivity analysis, parameter estimation, and uncertainty analysis that reflects conditions in both model grids.

This report describes the method used by LGR, evaluates LGR accuracy and performance for two- and three-dimensional test cases, provides input instructions, and lists selected input and output files for an example problem. It also presents the Boundary Flow and Head (BFH) Package, which allows the child and parent models to be simulated independently using the boundary conditions obtained through the iterative process of LGR.

Introduction

Simulations of ground-water flow and transport often need highly refined grids in local areas of interest to improve simulation accuracy. For example, refined grids may be needed in (1) regions where hydraulic gradients change substantially over short distances, as would be common near pumping or injecting wells, rivers, drains, and focused recharge; (2) regions of site-scale contamination within a regional aquifer where simulations of plume movement are of interest; and (3) regions requiring detailed representation of heterogeneity, as may be required to

1

simulate faults, lithologic displacements caused by faulting, fractures, thin lenses, pinch outs of geologic units, and so on.

Refinement of the finite-difference grid used by MODFLOW can be achieved using globally refined grids, variably spaced grids, or locally refined grids.

Using a globally refined grid – a grid refined over the entire domain – can be computationally intensive. In some cases the execution times are so long that they result in a computationally intractable problem; in others unnecessarily long execution times interfere with model development and utility. In addition, it may be inconvenient and unnecessarily labor intensive to develop the data sets required to refine an entire grid when only a local area is of interest.

Using a variably spaced grid, a fine grid can be attained locally with more moderate increases in computational time, but often results in refinement in areas that do not need such detail. This arises because the finite-difference method requires that the same grid spacing extend out to the boundaries and has two important implications: (1) if refinement is needed in multiple areas of the domain, using a variable-spaced grid often results in a relatively fine grid over the entire domain, and (2) in addition to introducing surplus nodes and therefore more computations, this approach can produce finite-difference cells with a large aspect ratio, which can lead to numerical errors (de Marsily, 1986, p. 351).

Using a locally refined grid can be less computationally intensive than the other two methods. This method, termed local grid refinement (LGR) in this report, links two or more different-sized finite-difference grids: a coarse grid covering a large area which incorporates regional boundary conditions, and a fine grid covering a smaller area of interest. These grids are often called parent and child grids, respectively, and this terminology is used in this report. Grid refinement can be vertical as well as horizontal. The link between the parent and child grids can be accomplished as a so-called one-way coupling, in which conditions simulated by the parent grid are imposed on the boundary of the child grid. Alternatively, the link can be accomplished in a way that also includes feedback from the child grid to the parent grid, thus allowing two-way communication between the grids. Solutions with feedback can be achieved either through iteration or simultaneous solution schemes.

In the field of ground-water modeling, one-way coupling is commonly called telescopic mesh refinement (TMR) and is most commonly accomplished using some form of interpolation of either heads, or fluxes, or both, from the coarse grid onto the boundaries of the child grid (for example, Ward and others, 1987; Leake and Claar, 1999; Davison and Lerner, 2000; Hunt and others, 2001). This approach is fairly straightforward and works well for many problems. However, the one-way coupling does not allow for feedback from the child grid to the parent grid. Thus, after running both models, the burden is placed on the modeler to check if heads along and fluxes across the interfacing boundary are consistent for both models (Leake and Claar, 1999, p. 5-7). If they do not match, there is no formal mechanism for adjusting the models to achieve better agreement. In this way, TMR methods generally lack numerical rigor and are prone to significant, often undetected errors (Mehl and Hill, 2002 and examples 1 and 3 of this report).

A numerically rigorous method that ensures that heads and fluxes are consistent between the two grids is needed to obtain dependably accurate solutions. LGR as documented in this report uses an iteratively coupled method. Two-way iterative coupling is used to ensure that the models have consistent boundary conditions along their adjoining interface. The method implemented here couples the models using shared nodes. That is, the grids are constructed such that nodes of the parent grid are coincident with selected boundary nodes of the child grid.

Purpose and Scope

The purpose of this report is to document LGR (local grid refinement) for MODFLOW-2005 (Harbaugh, 2005).

This report first provides highlights of LGR and discusses its compatibilities with other MODFLOW-2005 capabilities. Next, the method of local grid refinement used by LGR is described in detail, and the accuracy and the convergence of the method are evaluated. Then, the performance of iterative grid refinement using LGR is compared to alternative methods for simple two- and three-dimensional problems and consequences of the results for field applications are discussed. Next, input instructions, and selected input and output files are provided in Appendices 1 and 2 for LGR and BFH, respectively. Finally, error propagation in LGR is illustrated in Appendix 3 and notes for MODTMR (Leake and Claar, 1999) users are provided in Appendix 4.

Acknowledgements

LGR was developed with support from the U.S. Geological Survey (USGS) Office of Ground Water, the USGS Ground-Water Resources Program, and the U.S. Department of Energy through the USGS Yucca Mountain Project. Arlen Harbaugh designed the data storage conventions of MODFLOW-2005 to allow for multiple grids. His careful organization was instrumental for making LGR possible. We are grateful for the reviews and comments provided by Chris Langevin of the USGS Florida Water Science Center and Jesse Dickinson and Stan Leake of the USGS Arizona Water Science Center which greatly improved the quality of this report.

Highlights and Compatibility

This section presents highlights important to those deciding on whether to use LGR. Advantages of LGR and user concerns for designing models that are compatible with LGR are listed here for user convenience; additional discussion of these points is presented in the report, as noted.

Highlights for New Users and Quick Reference

Highlights of LGR are organized into four topics: (1) accuracy, (2) execution time, (3) model setup, and (4) grid and time step design. LGR can be run to perform on-way coupling by setting MXLGRITER=1; the comments here apply when LGR is run iteratively (MXLGRITER > 1). MXLGRITER is defined in Appendix 1.

Accuracy

1) Local refinement can provide much of the improved accuracy achievable by global refinement with much smaller execution times (see Table 4, Table 5, and Figure 26).

2) Local refinement generally improves the accuracy of all parts of the simulated system. (see Parent Grid Error and Child Grid Error sections).

3) The greatest refinement ratio does not necessarily produce the most accurate solution (see Effects of the Refinement Ratio section).

4) Local grid refinement maintains the rate of convergence of globally refined grids for homogeneous and heterogeneous models. This means locally refined grids reduce error in a way that is consistent with global refinement, and supports local refinement as a valid alternative to global refinement (see Convergence Properties section).

Execution Time

1) LGR uses a solution method that iterates between the parent model and the child model. A single iteration requires one parent-model solution (execution time T_{parent}) and one child-model solution (execution time T_{child}), so execution time per iteration is approximately $T_{parent} + T_{child}$. The number of iterations varies depending on the heterogeneity and the grid discretization. Generally, between 10 to 20 iterations, are sufficient for most problems (see Convergence Properties section).

Model Setup

1) The parent and child models each require a MODFLOW Name file (Harbaugh and others, 2000, p. 7, 43) and associated set of input files. The unit numbers defined in these files need to be unique – a unit number used in the parent-grid model input and output files cannot be used for the child-grid model.

2) In the Basic Package input file (Harbaugh and others, 2000, p. 50) for the child model, set IBOUND = IBFLG (see Input Instructions in Appendix 1) for cells that border the parent model. This is the perimeter of the child model. Except for the perimeter of the child model, do not use the values defined by IBFLG and -IBFLG anywhere in the IBOUND arrays of the child or parent models.

3) LGR currently (2005) needs to have sensitivities calculated and parameter estimation performed using universally applicable programs such as UCODE (Poeter and Hill, 1998), UCODE_2005 (Poeter and others, 2005), PEST (Doherty, 2004), or OSTRICH (Matott, 2005).

4) If the DE4 solver (Harbaugh, 1995) is used on the refined grid, the decomposition cannot be reused from a previous time step or internal iteration, even if the model is linear. Thus, use IFREQ = 3.

Grid and Time-step Design

1) For LGR currently (2005), only one block-shaped volume of local refinement can be simulated. This is not a restriction of the method, but of the implementation presented here.

2) The shared-node coupling used by LGR requires child-grid spacing that is an odd integer factor of the parent grid. For example, ratios of refinement of 1:1, 3:1, 5:1, 7:1, and so on can be simulated; 2:1, 4:1, 6:1, and so on cannot be simulated.

3) For vertical refinement, the top of the child grid needs to coincide with the top of the parent grid. However, vertical grid refinement does not need to start at the top because a vertical refinement ratio of 1:1 can be used. This can be useful when thick upper layers are desired for simulating water-table conditions (see The Top and Bottom of the Child Grid section).

4) For vertical refinement, the bottom of the child grid coincides with the finite-difference nodes in any layer of the parent-model grid except in the top and bottom layers of the parent grid. That is, the child model replaces one half of the thickness of the parent cell along its bottom. For vertical refinement within the bottom parent layer, the child grid needs to coincide with the bottom of the parent grid. For vertical refinement of parent models with a single model layer, the child model extends from the top to the bottom of the layer (see The Top and Bottom of the Child Grid section).

5) For transient simulations, the time-step size needs to be the same for both models. This is not a limitation of the method, but of the implementation presented here. It is

most easily accomplished by defining identical stress period lengths and time step variables (PERLEN, NSTP, and TSMULT) in the Discretization input file (Harbaugh and others, 2000, p. 45). This may require defining more stress periods than would otherwise be required for the parent-grid model, child-grid model, or both.

Compatibility with Other MODFLOW Packages and Processes

LGR is integrated into MODFLOW-2005, and most MODFLOW-2000 Packages of the ground-water flow process have been converted to MODFLOW-2005 (Harbaugh, 2005). LGR is designed to simulate the parent and child grids as two separate MODFLOW-2005 models and iterate between them until a balance of heads and fluxes along the interface between these models is achieved. Within each separate model, most MODFLOW-2005 packages can be used with no or little alteration. Because the models are separate, different packages can be used for each model. For example, the parent model may use the DE4 solver while the child model uses PCG. This flexibility of iterative local grid refinement is one of its major advantages.

The shared-node method used to couple the parent and child models results in truncated cells between the two models along sides and along the bottom of the child model in many situations (see Lateral Interface between Parent and Child Grids and The Top and Bottom of the Child Grid sections). For these interfaces, most of the interface cells are half cells. The interface cells have less cell volume and, in some coordinate directions, less area than cells in the remainder of the grid. For sinks, sources, and other stresses or boundary conditions that cross the interface between the models, the user may need to modify input for the cells where the grids are coupled, as described in the following section of this documentation. Table 1 lists the packages and processes of concern and identifies attributes that may need to be adjusted when using LGR.

Table 1. MODFLOW-2005 packages and processes that may need to be adjusted by the user.

Package or Process[1]	
Supported	**Comments**
BCF, LPF, HUF	Accuracy may be better if hydraulic properties are the same for adjoining cells at the interface
RIV, DRN, GHB	Modify conductances to account for the cell area at the interface
EVT, RCH	Modify rates to account for the cell area at the interface (see text)
DIS	The time steps need to be identical in the coupled models.
	The grid refinement ratio needs to be odd: 1:1, 3:1, 5:1, and so on.
DE4, PCG, SIP, LMG, GMG	When using the DE4 solver for the child grid, set IFREQ to 3. All other solvers require no change.
	Different solvers can be used for the parent and child grids.
Limited Support	**Description**
STR, MNW	These packages can be used within each grid, but routing of water across the interface between model grids is not supported.
Not Supported	**Available alternatives**
SEN Process	Perturbation sensitivities can be calculated using UCODE, UCODE_2005, PEST, OSTRICH, and so on
PES Process	Parameter estimation can be accomplished using UCODE, UCODE_2005, PEST, OSTRICH, and so on

[1]The three letter acronyms identify Ground-Water Flow Process Packages unless noted. See the list of acronyms preceding the abstract of this report for definitions

Input Files that May Need to be Changed For Cells at the Interfacing Boundary

In the shared-node method of local grid refinement, data input may require modification for the cells that form the interface between the parent-child grids if the DRN, EVT, GHB, RCH, RIV, or WEL Packages are used at these cells. Except for the WEL Package, the flux calculated by these packages depends on the full cell area, either directly (RCH and EVT) or through a user input conductance term (DRN, GHB, and RIV). Cells at the interface of the parent model are truncated, but in the packages listed the cell areas are not modified. Thus, the user needs to adjust inputs so that the proper influence of the stress is represented in the model. The original cell (not truncated) and four options are illustrated in Figure 1a-e using uniformly distributed recharge as the stress. The child grid uses specified-head boundary conditions at the interface, and at these cells, head-dependent and flux boundary conditions are not accounted for in the equations solved by MODFLOW, and therefore these contributions do not appear in the budgets for the child model. The user has four options for handling stresses at the interface cells:

1) No change (Figure 1b). This is not recommended because the effected of the stress in the parent grid is double counted because the area of the original cell is used.

2) Account for the truncated parent cells at the interface, but neglect the lost volume of stress at the specified-head boundary of the child interface. This may be a good assumption if the neglected stress is small relative to the remaining stress in the system. This option is depicted in Figure 1c.

3) Account for the truncated parent cells at the interface and account for the stress in the area of the child interface by modifying the stress in the adjacent parent cell. This option is depicted in Figure 1d.

4) Account for the truncated parent cells at the interface and account for the stress in the area of the child interface by modifying the stress in the adjacent child cell. This option is depicted in Figure 1e.

While options three and four may seem odd, limited testing suggests that they produce results that more closely match globally refined model results. If the third or fourth option is used, the conductances, stresses, or stress rates can be adjusted in either the adjacent parent or child cells to account for the neglected flux at the interface specified-head cells. The values should be adjusted such that the total volume (in the case of recharge) or total area (in the case of conductance-based stresses) is accounted for. For example, the volume of recharge at the interface specified-head cells can be accounted for by adjusting the recharge rates in either the adjacent parent or child cells so that this volume of flux is still represented in the model, as shown in Figure 1d and Figure 1e, respectively. If the parent cells are adjusted for uniform, areally distributed stresses, such as uniform recharge, the rate in the parent interface cells can be adjusted using equation 1a (and eqn. 1b for interface cells at the corners) to account for the additional stress over the area of the child interface cells. This is shown in Figure 1d. For the same stress, if adjustments are made to the child cells, the rate would be increased according to equations 1c and 1d for interface and corner cells, respectively. This is depicted in Figure 1e.

$$R_{adjustedParent} = R \times 0.5\ (1 + 1/NCPP) \tag{1a}$$

$$R_{adjustedParentC} = R \times [0.75 + 0.5 \times 1/NCPP - 0.25 \times 1/(NCPP)^2] \tag{1b}$$

$$R_{adjustedChild} = R \times 1.5 \tag{1c}$$

$$R_{adjustedChildC} = R \times 2.25 \tag{1d}$$

where

R = the original rate over the entire original parent cell area

NCPP = the number of child cells that span the width of a single parent cell.

$R_{adjustedParent}$ = the adjusted rate in the interface cell of the parent model

$R_{adjustedParentC}$ = the adjusted rate in the interface corner cell of the parent model

$R_{adjustedChild}$ = the adjusted rate in the interface cell of the child model

$R_{adjustedChildC}$ = the adjusted rate in the interface corner cell of the child model

If higher spatial resolution of the stress data is available, the higher spatial resolution allowed by the child grid can be combined with this data to dictate how stresses are partitioned. These situations can be accommodated by adjusting values in the corresponding cells such that the total influence of the stress is accounted for.

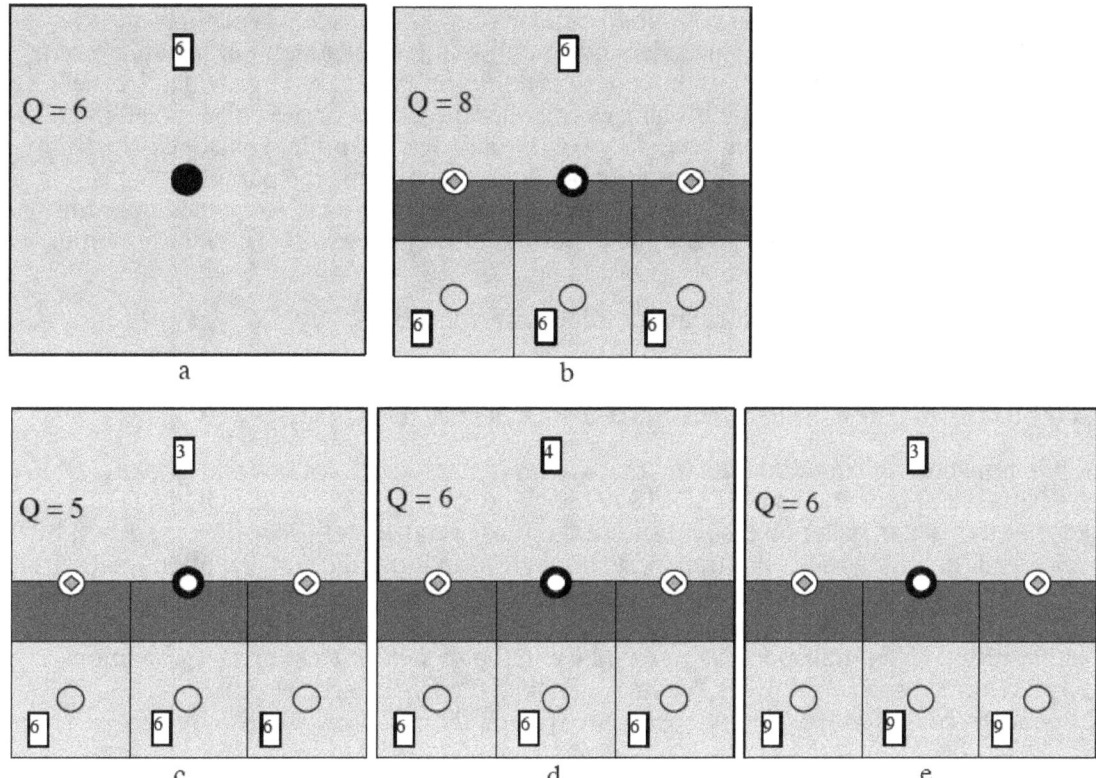

EXPLANATION

■   Area where specified-head designation means that in MODFLOW the recharge is not included in the equations solved

☐   Area where stress is accounted for

5   Recharge rate specified in the RCH Package input file

Q   Value of net recharge to the entire area assuming unit width and length for the area

Figure 1. Schematic of recharge for (a) the original uniform recharge distribution in the parent grid, (b) the recharge distribution on the interface with the original recharge rate used in the parent and child cells, which produces too large of a net recharge; (c) the recharge distribution with modifications to the parent model to account for the truncated cell size at the interface, but neglecting the lost recharge at the child interface; (d) accounting for the truncated parent cell and adding the excluded child recharge to the adjacent parent cell; and (e) accounting for the truncated parent cell and adding the excluded child recharge to the adjacent child cells.

The parent grid uses specified-flux boundary conditions at the interface. If a constant-head boundary is specified in the parent model at the parent-child interface, the specified fluxes cannot be defined. In this case, the constant-head boundary could be approximated by a general-head boundary with an appropriately large value of conductance.

Running Parent and Child Models Independently Using the Boundary Flow and Head (BFH) Package

The parent and child models can be simulated independently by using the coupling flux and head boundary conditions produced by LGR. This can be accomplished using the new

Boundary Flow and Head (BFH) Package which reads the coupling boundary conditions saved by LGR. Running the models independently can be useful when simulating solute transport, particle tracking, or other processes which do not affect the coupling boundary conditions produced by LGR.

Situations that might affect the coupling boundary conditions, such as changes in pumpage, can also be simulated using an independent child or parent model, but results become invalid as the changes affect the coupling boundary conditions. An analysis provided by the BFH Package can be used to determine if changes to either the parent or the child model requires re-running LGR to update the coupling boundary conditions. For example, consider a situation in which after running LGR and finding coupling boundary conditions, the parent model is updated to include new pumping data for a well outside the refined area. How much does this well change the interfacing boundary where the child model is coupled? BFH Package output can be used to answer this question.

Instructions for the BFH package are presented in Appendix 2.

Using LGR to Simulate Solute Transport and Particle Tracking

Solute transport and particle tracking that are limited to the parent or child grid are simulated easily. Programs such as MT3DMS (Zheng and Wang, 1999) and MODPATH (Pollock, 1994) that act as post processors can use the binary cell-to-cell flow files produced by each model for LGR simulations. For transport simulations that use the flow solution internally, such as ADV2 (Anderman and Hill, 2001) and the GWT Process, independent child or parent simulations can be run. In this case, first a LGR simulation is used to produce the coupling boundary conditions. Then transport is simulated using the models with boundary conditions provided by LGR and used by the BFH Package. That is, the BFH Package provides the coupling boundary conditions for the independently run model.

Solute transport and particle tracking that cross from one model to another are more difficult, but can be accomplished by using the results of one model as the boundary conditions for the other model. In general, particle tracking through cells that contain sinks/sources is problematic (see Pollock, 1994, p.2-14), which is the case for the interface cells. Therefore, tracking particles across the interface is an approximation. Particles that are transported across the boundary of a grid interface can be approximated manually by recording particle times and locations as they leave one grid and using those times and locations to begin transport of the particles in the adjacent grid. LGR as presented in this report has no method of translating particles or simulated concentrations across the interface. This can be accomplished by simulating the two models independently and doing the translation manually.

Solute transport across interfacing grid boundaries is difficult to represent accurately because the abrupt change in grid size. In addition, often the parent grid size is large enough that substantial numerical dispersion would be expected. Ideally, the grids should be designed such that important features of any solute plume remain entirely within a single grid.

Description of Local Grid Refinement (LGR)

The function of the child model is to simulate phenomena that need a finer grid than the parent-model grid. For example, relatively fine grids are often needed to represent accurately sharp changes in hydraulic gradient, abrupt changes in hydraulic properties that would otherwise lose resolution if represented by the coarser parent grid, or other processes such as solute transport for which a fine grid is often needed to obtain accurate solutions. The role of the parent model is to provide the boundary conditions to the child model that are consistent with the more regional flow system.

Description of Local Grid Refinement

LGR uses the iteratively coupled shared-node method of local grid refinement developed and tested by Mehl and Hill (2002 a,b; 2003; 2004) and Mehl (2003). The basic flow of the LGR procedure is shown in Figure 2 and the basic program flow of LGR within MODFLOW-2005 is shown in Figure 3. The components of the iterative coupling are discussed in the following sections.

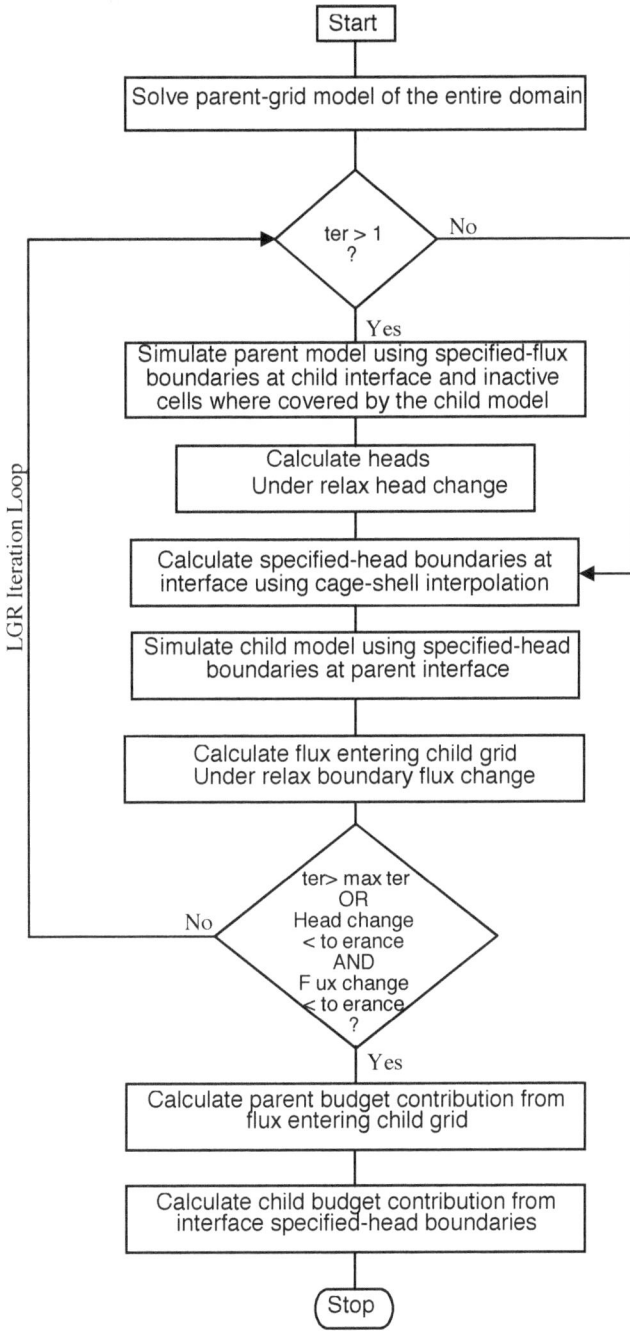

Figure 2. Flow chart for the iteratively coupled LGR procedure.

9

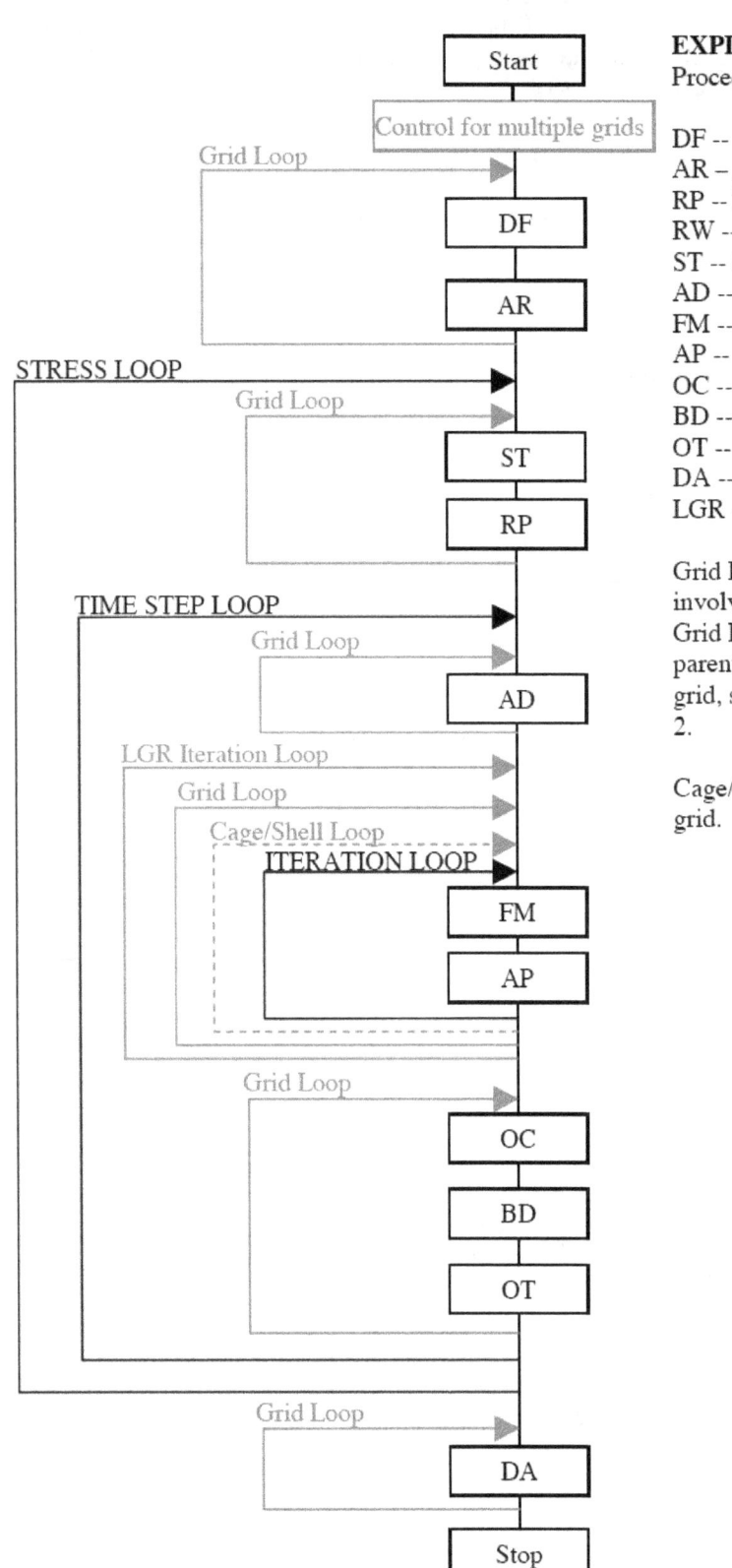

Figure 3. Flowchart of MODFLOW with Grid Refinement.

The Grid Structure of the LGR Shared-Node Method – Parent Grid, Child Grid, and the Interface

The grid structure is defined by how the sides, top, and bottom of the child model are nested within the parent model. The two models join along what are called interfaces. The lateral boundaries of the child model are always interfaces with the parent model. The top boundary of the child model is always the top of the simulated saturated ground-water system of the parent model. The bottom boundary of the child model may coincide with the bottom of the simulated system or may be an interface with the parent model.

Lateral Interface between the Parent and Child Grids

The lateral interface forms the sides of the child model grid. A schematic through the center of one layer of a locally refined grid is shown in Figure 4a. The interior cells of the parent model covered by the child model are made inactive by LGR by setting IBOUND to zero for these parent-model cells. Thus, after an initial parent-grid solution, the parent model has a hole in it that is filled by the child model. The parent and child models do not overlap cell areas. The model cells along the parent-child interface end at the node. Some nodes are shared and this gives the shared-node method its name. Most cells along both the parent and child side of the interface are half cells. That is, they represent half the volume that they normally would based on their cell dimensions. At the corners, the child grid has 1/4 cells and the parent grid has 3/4 cells. If the refinement does not extend to the bottom of the parent model (see next section), the child corner cells in the bottom layer are 1/8 cells and the parent has 7/8 cells.

To obtain ground-water flow equations that account for all aquifer material in each direction once, the conductances parallel to the parent-child interface are multiplied by 1/2, 1/4, or 3/4 and the storages for cells along the interface are multiplied by 1/2, 1/4, 3/4, 1/8, or 7/8. This arrangement produces grids that do not protrude into each other.

Shared nodes are obtained by using a parent-grid spacing that is an odd integer factor of the corresponding child-grid spacing. For example, in the grid presented in Figure 4, three child cells span the width of one parent cell, producing what is referred to as a 3:1 refinement ratio.

For the program described in this report, the refinement along rows and columns needs to be the same in the two directions for all rows and columns. This is not a requirement of the method, just a characteristic of this implementation.

The Top and Bottom of the Child Grid and Vertical Refinement

The top of the child model needs to coincide with the top of the simulated system of the parent model. The bottom of the child model can coincide with the bottom of the parent model or any nodes of any parent-model layer except nodes of the top or bottom layer. Figure 5 shows which vertical refinement scenarios are possible and which are not possible.

For single-layer parent models (Figure 5a), a vertically refined child model can be used. In this case, the interface nodes above and below the shared node are set to the value of head at the shared node. This means no vertical gradients are simulated along the interface.

The uppermost node(s) of the child model grid are between the top of the model and the uppermost nodes of the parent model grid. In this case, the interface node(s) directly above the uppermost shared nodes are set to the value of head at the shared node. This is consistent with a parent grid that does not have vertical flow between the top of the model and the uppermost node. Vertical flows from recharge or discharge will not, therefore, be correctly produced in the child grid at this location.

Vertical refinement can vary layer by layer. For example, extra refinement at the top is illustrated in Figure 5b where the top parent layer is refined vertically 5:1 and the second parent layer is refined vertically 3:1. Figure 5c shows that although the child refinement begins at the top layer, a 1:1 ratio can be used. Having thicker upper layers may be helpful for problems with rewetting (see discussion in Unconfined Conditions section).

If the child model extends to the bottom of the parent model, the interface node(s) that are directly below the bottommost shared node are set to the value of head at the shared node above. This is consistent with a parent model that does not have vertical flux between the bottom of the system and the bottommost node.

Figure 5d shows that it is not possible for the child model to terminate at the shared node of the top layer of the parent model. The child model must extend at least to the second layer of the parent model, as shown in Figure 5b. Figure 5e illustrates that it is not possible for the grid of the child model to terminate at the shared node of the bottom layer of the parent model. The child grid can extend past the bottommost shared node to the aquifer bottom, as shown in Figure 5c.

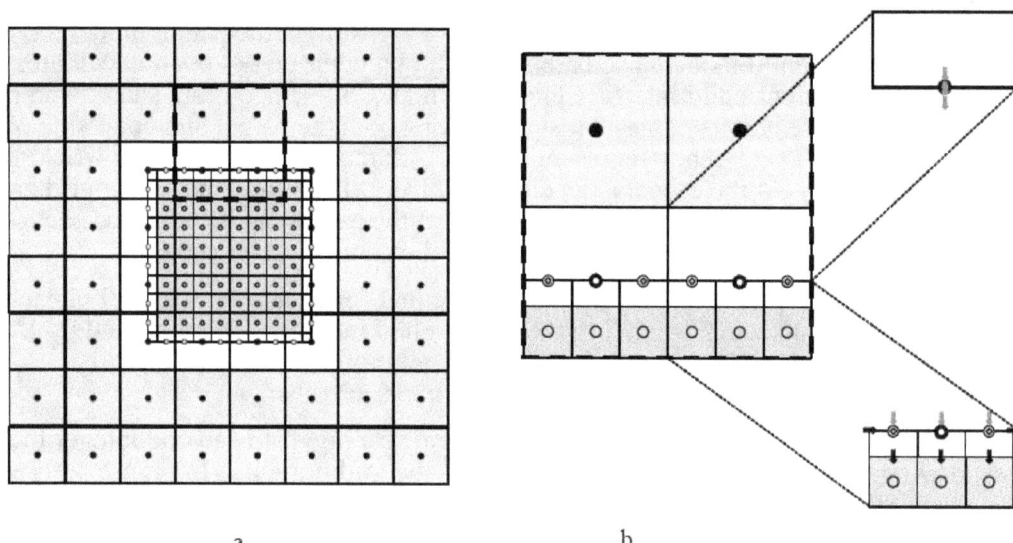

a   b

EXPLANATION

- ●   Node of the parent model only
- O   Shared node used by both the parent model and the child model
- ○   Node of the child model only. The parent model is inactivated here after the initial parent simulation, so the parent model has a hole in it.
- ◎   Specified-head boundary node of the child model determined by interpolation from the parent solution at the shared nodes
- ▮   Internal child-grid fluxes
- ▮   Fluxes summed to provide parent-flux boundary condition

Figure 4. (a) Two-dimensional areal schematic through the center of the locally refined grid. The interface area denoted by the dashed rectangle is shown in greater detail in (b), and illustrates flux balance across the parent-child interface. Cells fully within the child grid are darkly shaded, cells fully within the parent grid are lightly shaded, and cells at the interface are white.

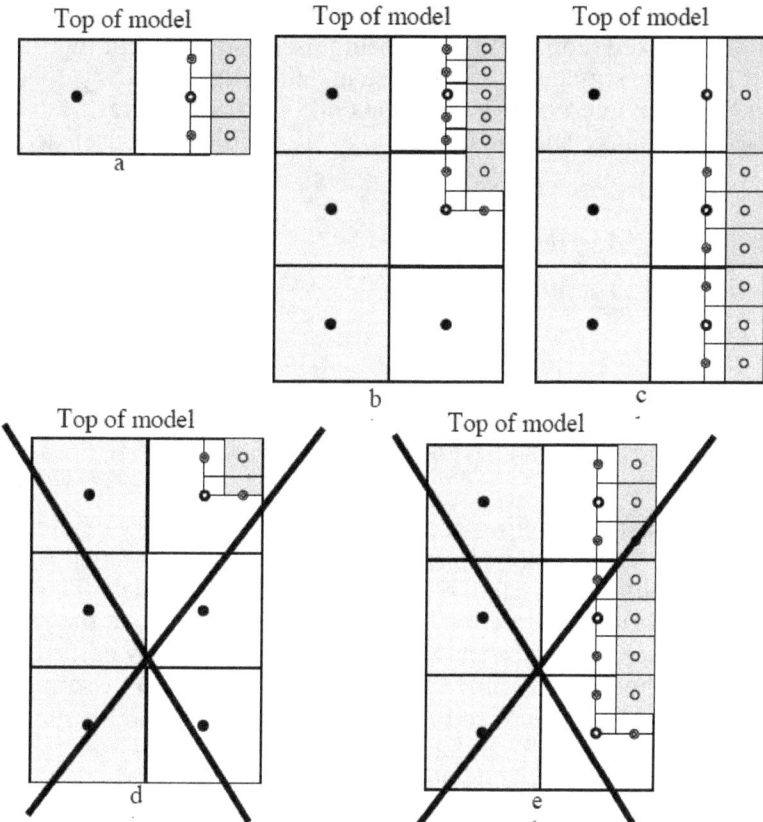

Figure 5. Cross-sectional schematic of vertical refinement interface of (a) a one-layer parent model refined to a three-layer child model, (b) a multi-layer parent model where the child refinement varies vertically and terminates at the shared node of the second parent layer, (c) a multi-layer parent model where the child refinement varies vertically and extends to the bottom of the parent model, (d) a multi-layer parent model where the child refinement terminates at the first shared node of the parent, which is not possible, and (e) a multilayer parent model where the child refinement terminates at the bottommost shared node, which is not possible.

The Iterative Coupling

The iteratively coupled shared-node method of local grid refinement balances heads and fluxes across the interfacing boundary of the two grids. This is accomplished by iteratively updating the head (child grid) and flux (parent grid) boundary conditions along the interface for each model. Relaxing (averaging) with the head and flux values from the previous iteration is needed to keep the iterations stable. This approach of coupling the two grids is similar to what is used by domain decomposition methods (DDM). However, most DDM operate at the matrix level – they formulate the matrix equations first, and then break up the matrix equations into separate problems. LGR operates at the ground-water system level – the ground-water system is divided into parent and child grids, and then the equations for each are formulated. Similar approaches have been used by Funaro and others (1988) and Nacul (1991), which operate at the partial differential equation and reservoir level, respectively.

As shown in Figure 2, the LGR procedure begins by simulating a parent model that encompasses the entire domain. For subsequent iterations the parent-model cells completely covered by the child grid are eliminated. For the cells along the interface (Figure 4), the conductances parallel to the interface and aquifer storages are adjusted. The heads from the parent model are used to interpolate specified-head boundary conditions for the child grid. The interpolated heads are relaxed using heads from the previous iteration (eqn. 2a). This under relaxation is necessary for convergence (Funaro and others, 1988; Székely, 1998). The value of the relaxation parameter is problem dependent and may have to be adjusted to achieve convergence. The child model is simulated and the fluxes through the parent-child interface are calculated and also under relaxed (eqn. 2b) before being used as the parent interior flux boundary condition. The parent model is simulated using these updated flux boundary conditions and produces updated heads for the interpolation onto the child grid boundaries. This process is repeated until both the head change and the flux change are smaller than user-defined criteria.

$$\text{head}^{\text{updated}} = \omega \cdot \text{head}^{\text{new}} + (1-\omega) \cdot \text{head}^{\text{old}} \qquad (2a)$$

$$\text{flux}^{\text{updated}} = \omega \cdot \text{flux}^{\text{new}} + (1-\omega) \cdot \text{flux}^{\text{old}} \qquad (2b)$$

where,

ω is the relaxation factor with values $0 < \omega < 1.0$

In the iterative method, the coupling occurs through the boundary conditions, which are accounted for in the right-hand side of the matrix equations. Thus, the stencil for the coefficient matrix is always consistent with the standard stencil of the original model. This is different from other two-way coupled local grid refinement methods where equations for the irregular connections across the interface of the parent and child grids are directly embedded into a single coefficient matrix, thus altering the conventional stencil (for example, Wasserman, 1987; Ewing and others, 1991; Edwards, 1999; Schaars and Kamps, 2001; Haefner and Boy, 2003). For MODFLOW, the coefficient matrix is formulated symmetrically and all non-zero terms are located on the matrix diagonal and six off diagonals (McDonald and Harbaugh, 1988, p. 12-2 – 12-4). Therefore, when using the iterative coupling considered in this work, efficient solvers that are based on a conventional finite-difference stencil on a Cartesian grid, such as the solvers distributed with MODFLOW, can be applied without modification. The resulting matrix equations are:

$$[A_p]\{h_p\} = \{f_p(h_c)\} \qquad (3a)$$

$$[A_c]\{h_c\} = \{f_c(h_p)\} \qquad (3b)$$

where,

[ ] denotes a matrix and { } denotes a vector,

$[A_p]$ is the coefficient matrix for the parent grid. It contains conductances and storage properties and has the same structure and coefficients as a conventional finite-difference discretization, except for adjusted conductances and storages along the interface with the child grid,

$[A_c]$ is the coefficient matrix for the child grid. It contains conductances and storage properties and has the same structure and coefficients as a conventional finite-difference discretization, except for adjusted conductances and storages along the interface with the parent grid,

$\{h_p\}$ is the head in the parent grid,

$\{h_c\}$ is the head in the child grid,

$\{f_p(h_c)\}$ is the right-hand side for the parent grid, and includes the flux boundary condition along the interface with the child grid. This flux is determined by a mass balance (Figure 4b) and is a function of the heads in the child grid from the previous iteration (h_c). It also contains the storage terms from the previous time step and all other boundary conditions and stresses within the parent model, and

14

{f_c(h_p)} is the right-hand side for the child grid, and includes specified-head boundary condition along the interface with the parent grid. This boundary condition is determined from the previous head solution on the parent grid (h_p) using interpolation. It also contains the storage terms from the previous time step and all other boundary conditions and stresses within the child model.

A common approach to handle asymmetric matrices is to iteratively solve them by using symmetric solvers and splitting the coefficient matrix such that the asymmetric terms are evaluated on the right-hand side. For example, the vertical flow calculation under dewatered conditions (McDonald and Harbaugh, 1988, p. 5-21 – 5-23) uses this type of splitting as does the LVDA capability of the HUF Package (Anderman and others, 2002). In this regard, the iterative method outlined above can be viewed as a matrix splitting of a directly embedded approach, where the coupling terms that are not involved in the conventional stencil are placed on the right-hand side. Head values from the previous iteration are used to evaluate these terms, and instead of being applied directly, they are under relaxed using the heads of the previous right-hand side term.

The details of how the parent and child boundary conditions are calculated are discussed in the sections below. First the child to parent coupling is described followed by the parent to child coupling.

Child to Parent Coupling – Specified-Flux Parent Boundary Conditions

The specified-flux boundary condition along the parent-grid interface is calculated using a mass balance on the child cells that border each parent cell. In Figure 4b, the dark arrows denote the internal child-grid fluxes that are needed to calculate the net flow across the interfacing boundary. The net flow across the boundary equals the sum of the lighter vertical arrows. This net flow is used to define a specified-flux (Neuman) boundary condition for the parent model on the subsequent iteration.

Parent to Child Coupling – Specified-Head Child Boundary Conditions

For interface nodes that are shared with the parent model (Figure 4, dark circles with white centers), the heads calculated by the parent model apply directly. For the child nodes along the boundary that do not share the same location with a parent node (open circles with diamonds in the center), the head is interpolated. Linear or other low-order, geometrically based polynomial interpolation has been suggested (Quandalle and Besset, 1985; Ward and others, 1987; Ewing and others, 1991; Leake and others, 1998; Székely, 1998; Davison and Lerner, 2000). Figure 6 shows that in the presence of heterogeneity, linear interpolation produces heads that do not obey the physics of ground-water flow. Other geometric interpolation methods share this difficulty. For this reason, an alternative Darcy-weighted interpolation method is used in LGR that circumvents this problem. The interpolation is described in the following paragraphs. First, the concepts are illustrated analytically using a one-dimensional interface boundary of a two-dimensional model. The numerical implementation used by LGR is then described. The numerical implementation produces the same results as the analytical formulation for a two-dimensional parent model, and also apples when the parent mode is three dimensional.

Description of Local Grid Refinement

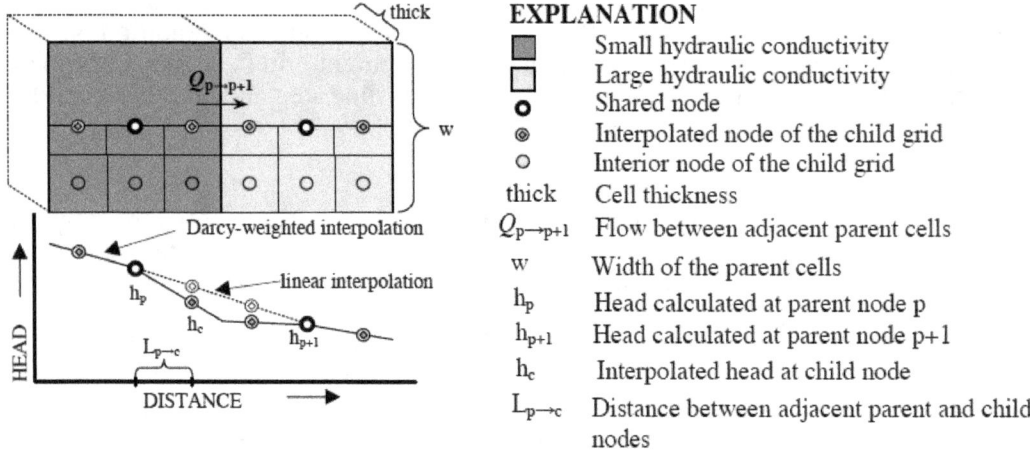

EXPLANATION

▨	Small hydraulic conductivity
☐	Large hydraulic conductivity
⊙	Shared node
⊚	Interpolated node of the child grid
○	Interior node of the child grid
thick	Cell thickness
$Q_{p \to p+1}$	Flow between adjacent parent cells
w	Width of the parent cells
h_p	Head calculated at parent node p
h_{p+1}	Head calculated at parent node p+1
h_c	Interpolated head at child node
$L_{p \to c}$	Distance between adjacent parent and child nodes

Figure 6. Darcy-weighted interpolation in relation to linear interpolation between shared nodes for cells with different hydraulic conductivity, as denoted by the different shading of the two cells.

Interpolation Concepts Illustrated Analytically Using a Two-Dimensional Model

The fundamental constitutive relation that governs heads and fluxes in ground-water systems is Darcy's law. In one dimension:

$$q = -K(dh/dx) \tag{4}$$

where,

q is the flux (Flow rate per unit area or Darcy velocity),

K is the hydraulic conductivity, and

dh/dx is the hydraulic gradient.

Darcy's law implies that if the one-dimensional flux, the material properties, and the physical dimensions are all known between two points, the hydraulic gradient at each location between the two points can be uniquely determined. Interpolating the heads along the boundary of the child grid using Darcy's law produces heads that are consistent with the parent-grid flow field. This interpolation also is consistent with finite-difference discretization of the ground-water flow equations. The resulting interpolation scheme, referred to here as Darcy-weighted interpolation, is calculated as:

$$h_c = h_p - \left(\frac{Q_{p \to p+1}}{K_c} \cdot \frac{L_{p \to c}}{thick \cdot \frac{w}{2}} \right) \tag{5}$$

where,

h_c is the head at the child node,

h_p is the head at the parent node,

$Q_{p \to p+1}$ is the flow between adjacent parent cells,

16

K_c is the hydraulic conductivity of the child cell, which needs to equal the hydraulic conductivity of parent cell for the most accurate interpolation,

thick is the thickness of the parent cell,

w/2 is half the width of the parent cell, and

$L_{p \to c}$ is the distance between parent node and child node.

Using w/2 is consistent with using the half conductances for the interface cells, as described previously. The Darcy-weighted interpolation is most accurate if the hydraulic conductivities of the child cells along the boundary are the same as those of the adjoining parent cells.

Various forms of one-dimensional Darcy-based interpolation have been developed and used by others (Wasserman, 1987; Schaars and Kamps, 2001; Haefner and Boy, 2003). For confined flow, they all produce the same interpolated heads, and these heads are consistent with the flow of the parent grid. However, they are not readily extendable to the two-dimensional interfaces of three-dimensional models.

Generally Applicable Numerical Interpolation Procedure

The method described above can be used to form a basis for the extension to three dimensions. First consider a numerical alternative to the analytical approach (eqn. 5) for interpolating heads onto the boundary of the child model used in two dimensions. The same interpolation could have been achieved by solving a one-dimensional numerical flow model for head at the child nodes between the shared nodes. In such a model, only the cells on the perimeter of the two-dimensional child model of Figure 4 (the interface with the parent grid) would be used; all the cells within the child model would be inactive. The shared nodes would be specified head (using the values from the parent), and the heads calculated at the child nodes in between the shared nodes would have the same value as those obtained analytically.

In the equivalent procedure in three dimensions, the shared nodes would be set as specified head and heads at the child cells on the interface would be solved numerically; all other cells would be inactive. The resulting equations for the child boundary conditions, which are a subset of those in equation 3b, are:

$$[A_{cB}]\{h_{cB}\} = \{f_{cB}(h_p)\} \tag{6}$$

where,

[A_{cB}] is the coefficient matrix for the child grid cells along the boundary interface with the parent grid. All other child cells are inactive and are eliminated from the equations,

$\{h_{cB}\}$ is the head on the child grid boundary interface, and

$\{f_{cB}(h_p)\}$ is the right-hand side for the child grid cells along the boundary interface with the parent grid. It contains specified-head boundary conditions at the shared nodes using values from the parent-grid simulation.

Unfortunately, when this one-step procedure is used to calculate the specified-head boundary conditions of the child model, it produces flows between child-grid nodes that are inconsistent with the flow between nodes of the parent grid (Mehl and Hill, 2004). To eliminate this problem, the following two-step interpolation procedure was developed.

Cage solution – The child nodes on the interface that are horizontally and vertically aligned with the shared nodes are calculated with specified heads at all the shared nodes. All other child nodes on the interface boundary are ignored. This is equivalent to one-dimensional solutions between the shared nodes and preserves the simulated flows between nodes of the parent grid. Figure 7a shows the nodes involved in this step for a part of the interface.

Shell solution – The shared nodes and the cage-solution nodes of step 1 are set as specified heads. Head at the remaining child nodes on the interface are calculated. This is equivalent to a series of two-dimensional numerical ground-water flow models. Figure 7b shows the nodes involved in this step for a part of the interface.

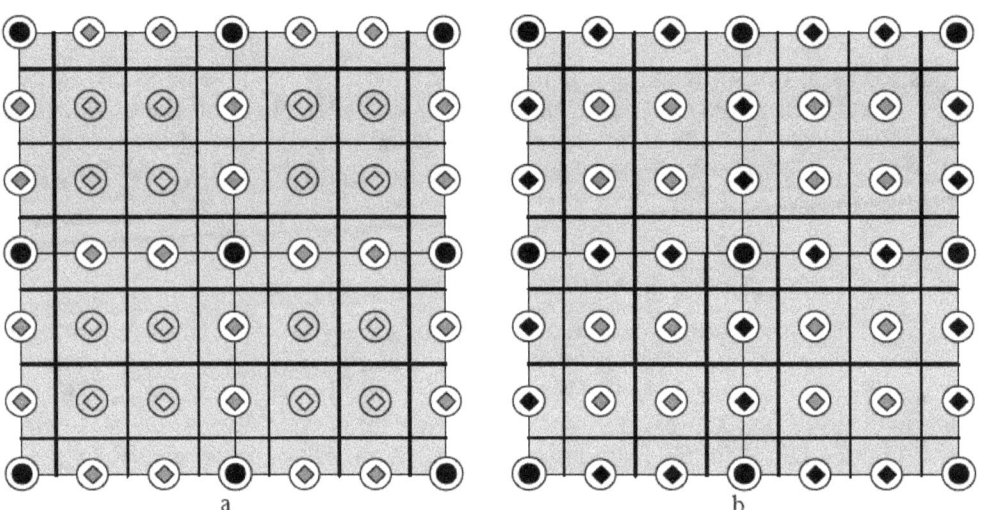

a   b

EXPLANATION

● Shared node (parent node that coincides with a boundary node of the child grid)

◈ Child node for which head is numerically solved in this step

◈ Child node that is inactive (ignored) in this step

◆ Child node where a specified head is imposed in this step

Figure 7. Schematic of the nodes used in the (a) cage solution and (b) shell solution for the head interpolation along the interface between parent and child grids.

The resulting equations for the cage solution, which are a subset of those in equation 6, are:

$$[A_{cage}]\{h_{cage}\} = \{f_{cage}(h_p)\} \tag{7a}$$

and the equations for the shell solution are the same as those in equation 6, except for the right-hand side:

$$[A_{cB}]\{h_{cB}\} = \{f_{cB}(h_{cage})\} \tag{7b}$$

where,

$[A_{cage}]$ is the coefficient matrix for the child grid cells along the boundary that form the cage – that is, the child cells that are directly in line with the shared nodes, as shown in Figure 7a. All other child cells are inactive and are eliminated from the equations,

{h_{cage}} is the head on the child grid boundary interface for cells that form the cage,

{$f_{cage}(h_p)$} is the right-hand side for the child grid cells along the boundary that form the cage. It contains specified-head boundary condition at the shared nodes using values from the parent-grid simulation, and

{$f_{cB}(h_{cage})$} is the right-hand side for the child grid cells along the boundary interface with the parent grid. It contains specified-head boundary conditions for the shared nodes from the parent solution and for the cells directly in line with the shared nodes (Figure 7b), using head values from the cage solution.

This two-step cage-shell interpolation procedure produces steady-state heads and flows that are consistent with the parent grid.

Unconfined Conditions

An advantage of solving for the child boundary heads numerically rather than analytically is that analytical solutions are much more difficult to achieve for irregular geometries or nonlinear flow phenomena, such as unconfined flow. These complexities are readily accounted for in the suggested numerical cage-shell procedure. Because the interpolation is handled numerically, the head interpolation is calculated iteratively using Picard iterations in the same way that MODFLOW calculates heads in an unconfined aquifer. In this way, the head solution on the boundary of the child grid is consistent with the parent grid solution whether the aquifer is confined or unconfined.

The rewetting capability often is used when simulating unconfined aquifers. This can be problematic for the coupling of parent and child grids. For example, if a parent cell along the interface between the two grids goes dry, then it no longer has a meaningful value and this causes problems for the cage-shell interpolation because the shared node has a value of head equal to HDRY. Therefore, for the version of LGR documented in this report, the child grid needs to be constructed such that it will not be adjacent to areas of the parent grid that are likely to go dry.

Grid refinement does not need to start at the top because a vertical refinement ratio of 1:1 can be used in any parent layer. If the child refinement does not extend to the bottom parent layer, an odd refinement ratio greater than one needs to be used for the lowest parent layer that is refined (see Figure 5b). Using thicker upper layers in the child model can be useful in alleviating some of the drying and rewetting problems associated with thin layers at the top of the model.

Closure Criteria

Closure criteria are needed to determine when to stop the LGR iterative procedure in addition to any closure criteria needed by the solver package. Closure criteria of the LGR iterations control the accuracy of both the head and flux boundary conditions, and thus the quality of the overall LGR solution. Given these boundary conditions, closure criteria for the solver package used by MODFLOW controls the accuracy of the parent and child solutions.

Closure Criteria for LGR Iterations

Separate closure criteria are needed for the parent and child grids. For convergence of the parent grid, the maximum relative change of the coupling specified-flux boundary condition between successive iterations needs to be less than a user defined amount (see equation 8a). For convergence of the child grid, the maximum change of the coupling specified-head boundary condition between successive iterations needs to be less than a user defined amount (see equation 8b).

$$|flux^{i+1} - flux^i|/max(|flux^{i+1}|, 1.0) \qquad\qquad (8a)$$

$$|head^{i+1} - head^i| \qquad\qquad (8b)$$

where,

superscript 'i' indicates the LGR iteration, and

|·| indicates the absolute value.

After convergence of the LGR iterations, the budget error in the parent model should be examined. This mass balance error includes the fluxes along the parent-child interface which are calculated from the subsequent child-grid solution (see Figure 2). It is an indicator of the overall quality of the LGR solution because it shows how precisely the flow in the parent grid balances the flow in/out of the child grid through the interface boundary. If the mass balance is deemed too large, lower the closure criteria of the LGR iterations until an acceptable mass balance is achieved. More LGR iterations may be needed to achieve the lower closure criteria. Generally, the same guidelines that are often used for standard MODFLOW simulations can be applied here, so that a budget error less than 1 percent is adequate. Because of errors introduced from the abrupt change in grid size, some mass balance error may remain which cannot be attenuated with further iterations. Therefore, the LGR iterations also are stopped after the user-specified maximum number of LGR iterations is exceeded.

Conversely, if the LGR iterations do not converge, but the budget error of the parent grid is small, then the quality of the LGR solution is probably acceptable. In this case, the closure criteria generally can be increased such that convergence is achieved and the mass balance is still acceptable.

The volumetric budget of the child grid printed by MODFLOW-2005 includes the boundary fluxes through the specified-head boundaries as part of the constant-head budget term. However, it does not report flows between adjacent constant-head cells and it includes fluxes from all constant-head boundaries, not just those on the interface with the parent grid. To provide more information about the interface, the child model output includes a separate budget for the specified-head cells involved in the LGR coupling. This is printed below the overall volumetric budget and corresponds to the parent-flux boundary reported in the parent volumetric budget.

An example is shown in Appendix 1.

Solver Iterations

The available solvers for MODFLOW-2000 – SIP, (McDonald and Harbaugh, 1988), PCG2 (Hill, 1990), DE4 (Harbaugh, 1995), LMG (Mehl and Hill, 2001) and GMG (Wilson and Naff, 2004) – are compatible with LGR. The parent and child models can use different solvers. Generally, the closure criteria used for the solvers should be less than or equal to what is used for the LGR closure. For example, it does not make sense to try to solve the coupling boundary conditions to a precision of 10^{-5} when the overall head solution, as controlled by the solver closure criteria, is only accurate to 10^{-3}. HCLOSE, which is in all solvers but LMG, can be compared to the head closure for the LGR iteration. Only rough guidelines can be provided here, but generally it is better to be cautious and use strict closure criteria for the solver. If this results in excessive computer processing times, make adjustments to the solver closure criteria.

Transient Simulations

For transient simulations, the iterative process described is repeated for each time step of each stress period, as shown in the flow chart of Figure 3. Like the conductances, storage coefficients of the parent and child grids are adjusted to account for the truncated cells on the interface (Figure 4).

For transient flow, the cage-shell interpolation procedure does not maintain perfect consistency with the flow of the parent grid, even for simple one-dimensional transient flows. This is because transient flow phenomena are propagated differently through different grid sizes. This problem is not limited to locally refined grids. For example, consider a coarse grid and a variably spaced grid which has the same grid spacing as a coarse grid near the boundary but is 9 times smaller near the center. Consider one-dimensional transient flow caused by instantaneously lowering the head at the right boundary by 9 m and calculate the percent difference in heads between solutions simulated using the coarse and the variably spaced grid. Normalize by 9 m to obtain the difference per meter of head change at the boundary. The results are shown in Figure 8. The hydraulic conductivity and specific storage are 0.25 m/day and 2.0×10^{-5} m^{-1}, respectively. Figure 8 shows three characteristics: 1) the greatest difference is at early time near the location where the transient phenomenon occurs, 2) the error dissipates with distance from the location of transient phenomenon, and 3) the error dissipates as the system approaches steady state. This implies that the difference in propagation of transient flow caused by using different grid sizes will introduce an additional error in the coupling. Generally, this error decreases as distance between the interfacing boundary and the location of the transient phenomenon increases and as the system approaches steady state.

This error along the boundary has implications for volumetric budget calculations for large-scale regional models that are used as the parent grid. In such models, small changes in head due to coupling errors at the interface can result in large changes in storage. For example, consider a regional model with cell dimensions of 1000×1000 m, a specific yield of 0.2, and coupled to a local model using LGR with a head closure criterion set at 1.0×10^{-2} m. Heads along the boundary can have errors on the order of 1.0×10^{-3}. This error, when viewed as a head change, can cause changes in storage on the order of $1.0 \times 10^{-3} \times 1000 \times 1000 \times 0.2 = 200$ cubic meters, which may be a significant amount of the overall budget depending on the flow system. Changes in storage are calculated on a per time-step basis, as

$$\Delta S / \Delta t \tag{9}$$

where,

ΔS = the change in storage = $(h^{n+1} - h^n) \times S_y \times A_c$

Δt = the time-step size

h^n = head at time step n

S_y = specific yield

A_c = Planar area of the cell

In accordance with equation 9, this error in storage is increased by smaller time steps and attenuated with larger time steps. Furthermore, within a given stress period, changes in storage decrease as time progresses which reduces this error. Therefore, storage changes at early time steps within a stress period can contain a significant amount of error but have reliable accuracy at later time steps. Users should keep these issues in mind when examining changes in storage for large-scale regional models coupled to local-scale models using LGR.

Evaluation of LGR Convergence and Numerical Accuracy Using a Two-Dimensional Test Case
with Varying Levels of Heterogeneity

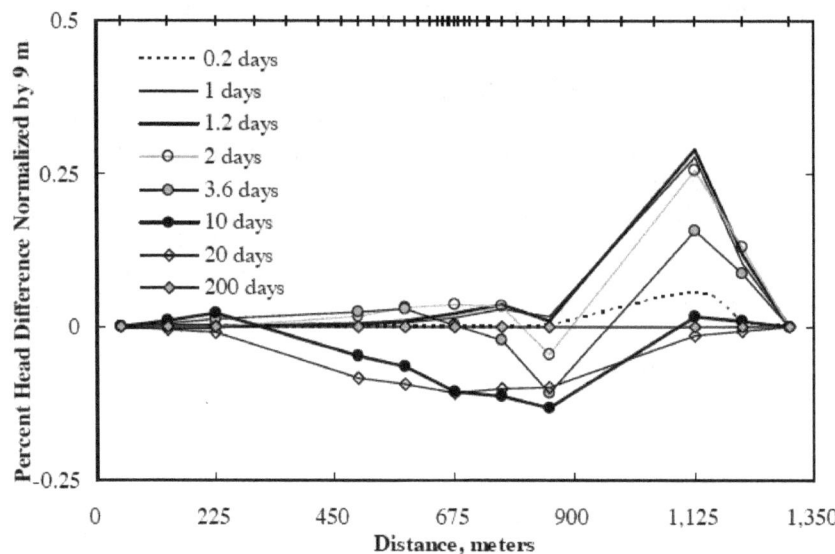

Figure 8. Percent difference in head normalized by 9 m at various times of a one-dimensional transient flow
simulated using a coarse grid and a variably spaced grid. The coarse grid spacing equals the largest spacing of the
variably spaced grid. Tick marks on the top axis indicate the grid spacing of the variably spaced grid. Transient flow
is caused by instantly lowering the head at the right boundary by 9 m while head at the left boundary remains fixed.

As currently (2005) implemented, LGR requires that the parent and child grids use
equivalent time discretization.

For transient problems, the child model cannot take advantage of the ability of the DE4
solver to reuse matrix decompositions (Harbaugh, 1995, p. 3). This is because the cage-shell
interpolation technique on the child grid changes the connections of the active cells (the cells
interior to the boundary cells are made inactive during the interpolation and then re-activated for
the full solution). Therefore IFREQ should be set to 3 for the child model so that the coefficient
matrix is reformulated for each solver iteration. The parent model simulation can still take
advantage of this part of the DE4 solver and IFREQ can be set to 1 or 2 when applicable.

Evaluation of LGR Convergence and Numerical Accuracy Using a Two-Dimensional Test Case with Varying Levels of Heterogeneity

A two dimensional test case is used to demonstrate the effects of parent-grid spacing and
the ratios of refinement on the accuracy and convergence of the local grid refinement procedure.
The test case features refinement around a pumping well using different ratios of refinement and
different levels of heterogeneity.

The test case system is shown in Figure 9. It is 1,260 m long and 1,350 m wide, has
constant-head boundaries of 10 m and 1.0 m on the left and right sides, respectively. No-flow
boundaries extend across the top and bottom. A pumping well extracts 5.5 m^3/s from the center
of the domain. The heterogeneity structure was constructed using five transmissivity (T) zones
randomly distributed throughout the domain with a spatial resolution equivalent to the coarsest
parent grid (90×90 m), as shown in Figure 9. This heterogeneity structure was selected to
provide a numerical challenge for LGR. Five levels of heterogeneity are considered. They begin
with a homogenous case (T=0.25 m^2/s for all zones). Then the contrasts of the transmissivities

represented by each zone are increased four times until the heterogeneity structure has a variance of Ln(T) of approximately 3.6 and a range of approximately three orders of magnitude.

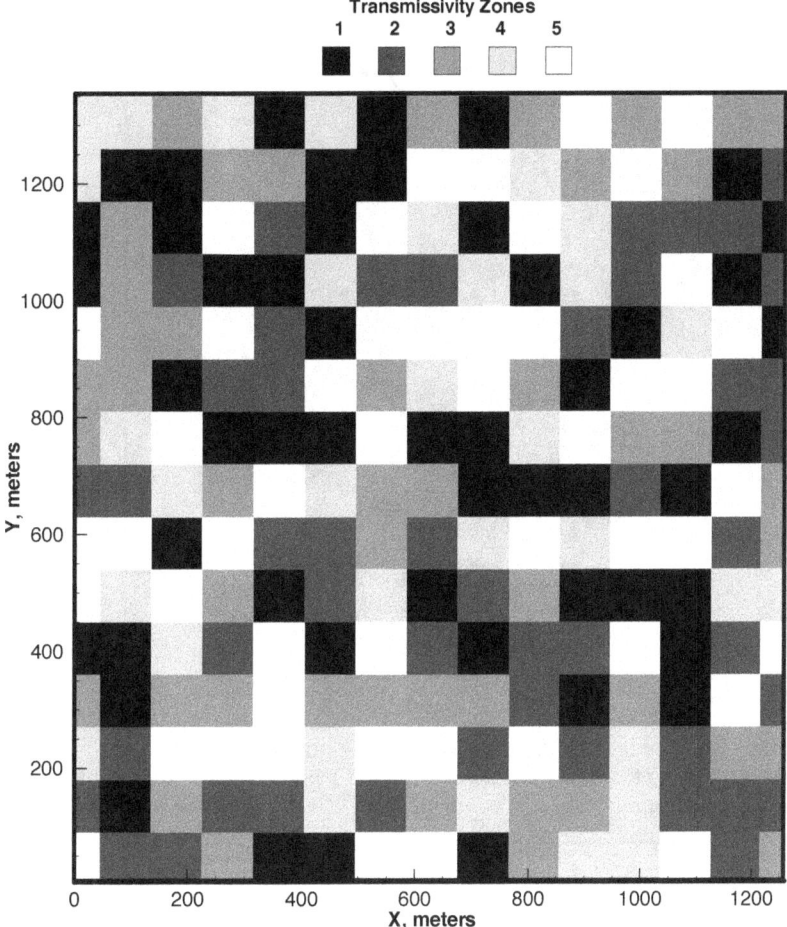

Figure 9. Heterogeneity structure with five transmissivity zones. Constant-head boundaries are imposed along the block-centered cells of the left and right sides.

Four different parent grid discretizations (ΔL) are used: 90×90 m, 30×30 m, 10×10 m, and 3.333333×3.333333 m. For each parent grid (except the coarsest and the finest), there are three child models, each spanning the same area in the vicinity of the well, each with a different ratio of refinement: 3:1, 5:1, and 9:1.

For all the simulations, the effectiveness of the local grid refinement procedure is evaluated by comparing 506 head locations throughout the grid to a "true" solution obtained using a uniform grid discretization of 1.111111×1.111111 m. The comparison locations are shown in Figure 10. Of the 506 head comparison locations, 146 correspond to the nodes of the coarsest parent grid, excluding the nodes along the interface with the child grid. The remaining 360 head comparison locations correspond to the nodes of the coarsest child grid, including the nodes along the interface with the parent grid, but excluding the node that corresponds to the pumping well. The comparison at the well node was excluded because when included, its error dominated all other errors and thus a good representation of the overall accuracy of the different grids was not achieved.

For the homogenous case, the fine grid "true" solution was compared to the analytical solution of Chan and others, (1976, equation 5). It was found that this grid discretization was adequate because the two solutions agreed within 5 to 6 significant digits. The analytical solution was not used for the comparisons because the infinite series is extremely slow to converge for observations that have the same y coordinate as the well and it cannot be used for

Evaluation of LGR Convergence and Numerical Accuracy Using a Two-Dimensional Test Case with Varying Levels of Heterogeneity

the heterogeneous cases. Figure 10 shows the head contours for the homogenous case; Figure 11 shows the head contours and flow field for the most heterogeneous case.

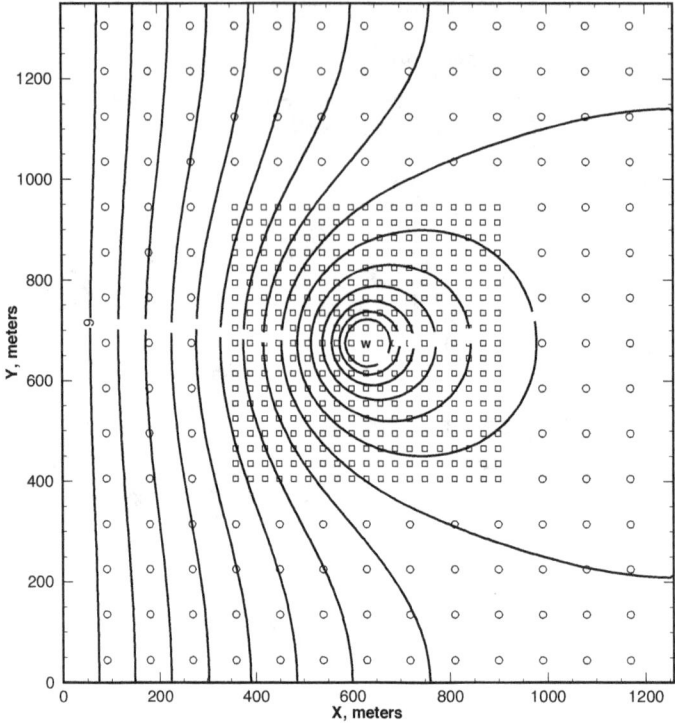

Figure 10. Head contours and head comparison locations for the homogenous case with a pumping well at the center. Constant-head boundaries at x=0 and 1,260 m and no-flux boundaries at y=0 and 1,350 m. "○" and "□" denote parent and child head comparison locations, respectively. "w" denotes the location of the pumping well.

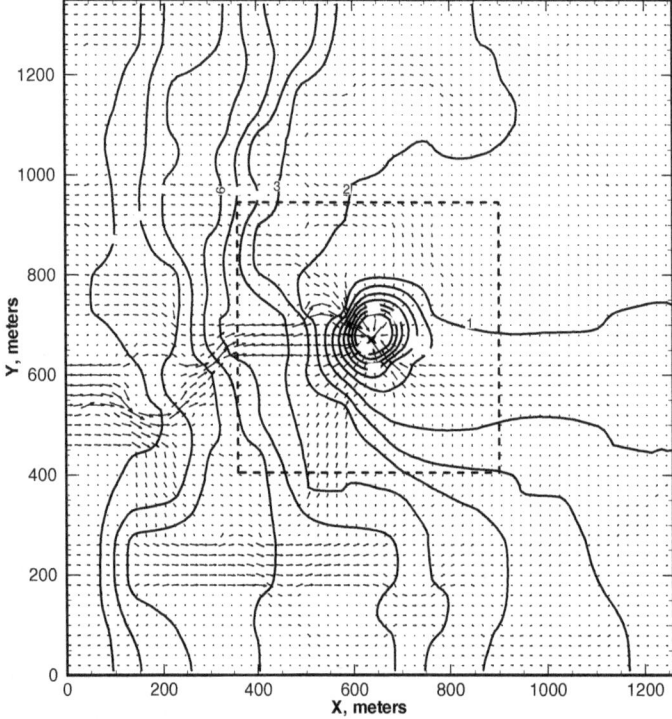

Figure 11. Flow vectors and head contours for the case with the highest variance of Ln(T). Area within the dashed rectangle indicates region of local grid refinement.

Evaluation of LGR Convergence and Numerical Accuracy Using a Two-Dimensional Test Case
with Varying Levels of Heterogeneity

Results are presented for the homogenous and most heterogeneous cases in Table 2 and Table 3, respectively. The tables show how the error, calculated as the mean of the L_1 norm (average of the absolute values) of the differences between heads in the coarser-gridded parent and child models and head in the fine grid model ("true" solution), is reduced using different levels and ratios of refinement. The errors in these examples are small, but close examination of the errors in these problems provides some guidance about how to best apply the method. This will be consequential for more difficult problems with larger errors, as presented later in this report as examples 1 and 3. The following sections discuss errors in the parent and child grids, the effects of the refinement ratio, and the convergence properties.

Parent Grid Error

The error in the parent grid improves by having the interior region of local grid refinement, as shown in Table 2 and Table 3. For example, in Table 2 the L_1 norm of errors using a uniform grid spacing of 90 m throughout was 5.714×10^{-3}, whereas that measure of error was 3.270×10^{-3} using a parent grid spacing of 90 m and a child grid spacing of 30 m (parent error for cell dimensions *90* and 90:30). This 43 percent reduction in error occurs because the feedback from the child grid, which has a finer discretization near the well and thus better characterizes the dynamics of the flow in the interior, positively influences the solution of the parent grid. Such improvements are not possible with a one-way coupled method of local grid refinement because there is no feedback from the child grid to influence the parent results.

Child Grid Error

Comparison of the uniform grid to the same size discretization of the child grid (for example, the child error of 2.782×10^{-2} and 9.594×10^{-3} for cell dimensions 90:30 and *30*, respectively in Table 2) show that the uniform grid has less error, as expected, but the child grid error is approximately within a factor of 3 of the uniform grid error. A similar analysis of Table 3 also shows that the error is approximately within a factor of 3 (1.261 versus 0.6373). This is a favorable result because it shows that reasonable accuracy can be obtained within a refined part of the domain without having to extend that same refinement throughout the entire domain.

The accuracy of the boundary conditions that the parent grid provides the child grid directly influences the child performance. This is demonstrated by comparison of errors within the child domain with the same discretizations, but different parent grids (for example, child error of 2.602×10^{-2} and 3.036×10^{-3} for cell dimension 90:10 and 30:10, respectively, in Table 2). This suggests that more accurate results can be obtained by using multiple generations of refinement (that is, a parent, child, grandchild, and so on) rather than a single, high-ratio refinement. These results are in agreement with those found by von Rosenberg (1982). LGR presented in this report does not provide for multi-generation refinement.

For this problem the child model domain errors (column 5 of Table 2 and Table 3) are larger than in the parent part of the domain (column 3). This reflects the sharper changes in gradients that occur in the child part of the domain, which are more difficult to model even with the finer discretization.

25

Evaluation of LGR Convergence and Numerical Accuracy Using a Two-Dimensional Test Case with Varying Levels of Heterogeneity

Table 2. Comparison of errors in simulated hydraulic heads for various grid levels and ratios of refinement using a homogeneous transmissivity distribution.

[Values in italics indicate results from a uniform-grid simulation; m, meter; -, values unavailable because nodes do not align; L_1 norm, average of absolute values]

Cell Dimension (m)		[1]Parent Grid Error (m)	Order of Parent Grid Convergence	[2]Child Grid Error (m)	Order of Child Grid Convergence	[3]Overall Error (m)	Order of Overall Convergence
Parent	Child						
90	-	*5.714E-03*	-	-	-	-	-
90	30	3.270E-03	-	2.782E-02	-	2.074E-02	-
90	10	3.290E-03	-	2.602E-02	-	1.946E-02	-
30	-	*6.186E-04*	*2.024*	*9.594E-03*	-	*7.004E-03*	-
30	10	3.770E-04	1.966	3.036E-03	2.016	2.269E-03	2.014
30	6	3.766E-04	-	2.688E-03	-	2.021E-03	-
30	3.333	3.779E-04	1.970	2.628E-03	2.087	1.979E-03	2.081
10	-	*6.782E-05*	*2.012*	*1.289E-03*	*1.827*	*9.370E-04*	*1.831*
10	3.333	4.208E-05	1.996	3.183E-04	2.053	2.386E-04	2.050
10	2	4.197E-05	1.997	2.928E-04	2.018	2.204E-04	2.017
10	1.111	4.209E-05	1.998	2.957E-04	1.988	2.226E-04	1.989
3.333	-	*6.840E-06*	*2.088*	*1.185E-04*	*2.172*	*8.632E-05*	*2.171*
3.333	1.111	4.265E-06	2.084	2.981E-05	2.155	2.244E-05	2.152
1.111	-	[4]*8.155E-07*	-	[4]*1.183E-05*	-	[4]*8.652E-06*	-

[1]L_1 norm of the errors associated with the comparison locations in the parent part of the domain.

[2]L_1 norm of the errors associated with the comparison locations in the child part of the domain.

[3]L_1 norm of the errors associated with all comparison locations in the domain.

[4]L_1 norm of the errors compared to the analytical solution, excluding comparison locations with the same y coordinate as the well.

Table 3. Comparison of errors in simulated hydraulic heads for various grid levels and ratios of refinement with a heterogeneous transmissivity distribution of Variance of Ln(T) ≈ 3.6.

[Values in italics indicate results from a uniform-grid simulation; m, meter; -, values unavailable because nodes do not align; L_1 norm, average of absolute values]

Cell Dimension (m)		[1]Parent Grid Error (m)	Order of Parent Grid Convergence	[2]Child Grid Error (m)	Order of Child Grid Convergence	[3]Overall Error (m)	Order of Overall Convergence
Parent	Child						
90	-	*5.096E-01*	-	-	-	-	-
90	30	5.079E-01	-	1.261E+00	-	1.044E+00	-
90	10	5.004E-01	-	1.179E+00	-	9.830E-01	-
30	-	*2.402E-01*	*0.685*	*6.373E-01*	-	*5.227E-01*	-
30	10	2.307E-01	0.718	5.618E-01	0.736	4.663E-01	0.733
30	6	2.277E-01	-	5.455E-01	-	4.538E-01	-
30	3.333	2.251E-01	0.727	5.335E-01	0.722	4.445E-01	0.722
10	-	*1.107E-01*	*0.705*	*2.796E-01*	*0.750*	*2.309E-01*	*0.744*
10	3.333	1.039E-01	0.726	2.515E-01	0.732	2.089E-01	0.731
10	2	1.019E-01	0.732	2.450E-01	0.729	2.037E-01	0.729
10	1.111	1.002E-01	0.736	2.401E-01	0.727	1.998E-01	0.728
3.333	-	*4.118E-02*	*0.900*	*1.011E-01*	*0.926*	*8.382E-02*	*0.922*
3.333	1.111	3.697E-02	0.940	8.940E-02	0.941	7.428E-02	0.941

[1]L_1 norm of the errors associated with the comparison locations in the parent part of the domain.

[2]L_1 norm of the errors associated with the comparison locations in the child part of the domain.

[3]L_1 norm of the errors associated with all comparison locations in the domain.

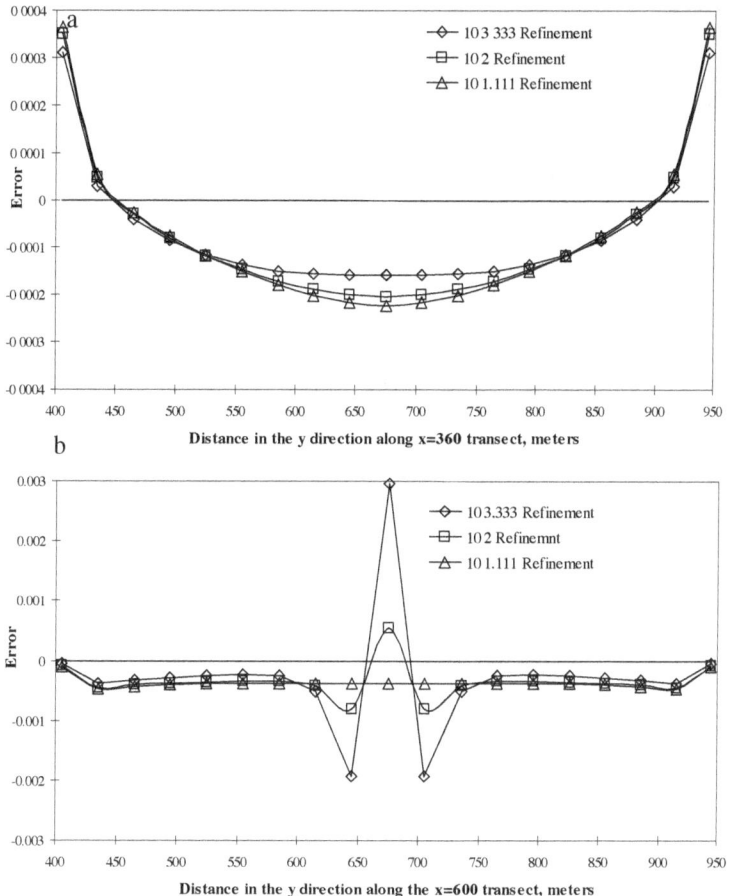

Figure 12. Distribution of error in the child grid along transects for the homogeneous model (a) along the interface with the parent grid at x = 360 meters and (b) next to the pumping well (x = 600 meters) for three different ratios of refinement.

Effects of the Refinement Ratio

Generally, the error within both the parent and child grid decreases as the ratio of refinement increases as suggested by comparing the errors of Table 2 and Table 3 for constant parent grid size and decreasing child grid size. However, errors at the boundary between the two grids may increase as the ratio of refinement increases. Figure 12 shows how the error of the child grid is distributed for the homogenous case along two transects; the first is along the boundary with the parent grid (Figure 12a) and the second is next to the pumping well (Figure 12b).

The results in Figure 12b indicate that higher ratios of refinement result in smaller error near the well, which is expected because the finer discretization associated with the higher refinement ratio is better able to characterize the influence of the well on the hydraulics. However, Figure 12a suggests that errors near the interface with the parent grid are larger for higher refinement ratios. These results are consistent with results found by Ewing and others (1991). These results illustrate that, for different refinement ratios, there is a tradeoff in accuracy near the well compared to the boundary; thus, there is an optimal level of refinement depending on what is important to the modeler. For example, for the homogenous case in Table 2, if the modeler is interested in overall accuracy, as opposed to accuracy just near the well, the 10:2 refinement, with an overall accuracy of 2.204×10^{-4} would be the best choice of the 10:x models presented.

27

Evaluation of LGR Convergence and Numerical Accuracy Using a Two-Dimensional Test Case with Varying Levels of Heterogeneity

Convergence Properties

The results for the homogenous problem (Table 2, columns 6 and 8) indicate that this method maintains the quadratic convergence associated with the standard centered-difference approximation. Additional simulations were conducted to investigate three intermediate levels of heterogeneity to produce the results shown in Figure 13. The convergence for the heterogeneous cases follow the convergence associated with a uniform grid refinement, as shown in Figure 13. These results demonstrate that as the hydraulic-conductivity field becomes more heterogeneous, the convergence decreases from quadratic. These findings might appear to be in contradiction to those found by Forsyth and Sammon (1988) and Weiser and Wheeler (1988), who show that the convergence is quadratic, even for heterogeneous cases. However, their results are based on the assumption that the hydraulic-conductivity field satisfies certain smoothness criteria; for example, the hydraulic conductivity monotonically increases in a given coordinate direction. Such criteria are clearly violated by the more complex random field shown in Figure 9. The consequences of these findings are significant for two reasons: 1) uniformly refining a grid will not reduce the error quadratically if the heterogeneity structure is not smooth, which is often the case in natural systems, and 2) the local grid refinement used in LGR does not diminish the model convergence properties.

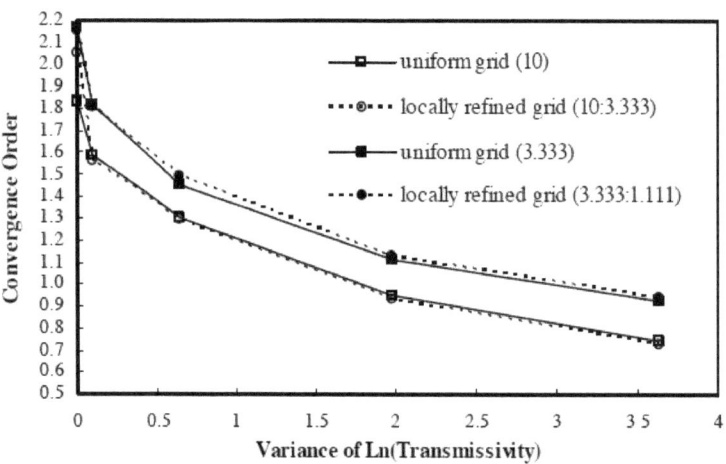

Figure 13. Convergence order in relation to variance or degree of heterogeneity, for uniformly and locally refined grids. Numbers in parenthesis indicate the grid spacing in meters.

The number of LGR iterations between the parent and child grids needed for convergence is important to the efficiency of the method. For the cases presented, the convergence criteria were met when the maximum fractional head and flux changes between iterations were less than 10^{-6}. Results for a subset of the simulations are shown in Figure 14 and demonstrate that the number of iterations needed for convergence increases with increasing heterogeneity in transmissivity.

Figure 14 also shows a variation in the number of iterations needed as the number of nodes increases. For the cases where variance of Ln(T) is 0 (homogenous) and 0.01, the number of iterations decreases as the number of nodes increases. However, as the variance increases, this trend is less pronounced (variance of 0.65). Eventually the opposite is true, and the number of iterations increases as the number of nodes increases (variance of 2.0 and 3.6). This is probably because the unrefined parent grid can adequately represent the hydraulics for the less heterogeneous cases: relatively few iterations are needed to converge because the feedback from the child is not providing much different information regarding the hydraulics. However, for the more heterogeneous cases where the hydraulics are more complicated, more iterations are needed to establish an equilibrium between the two grids.

Evaluation of LGR Convergence and Numerical Accuracy Using a Two-Dimensional Test Case
with Varying Levels of Heterogeneity

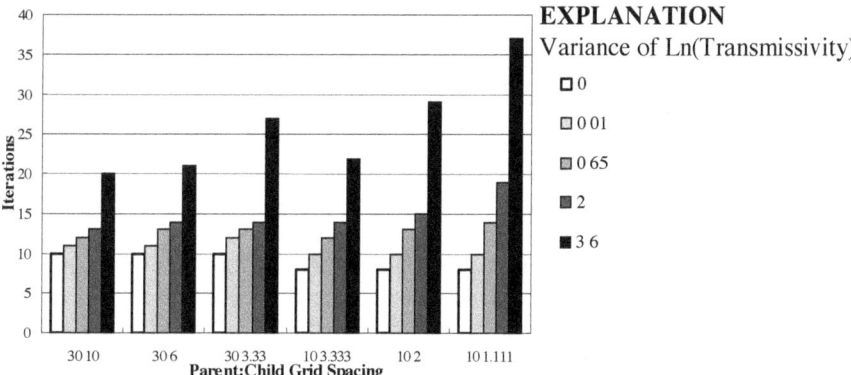

Figure 14. Number of local grid refinement (LGR) iterations between the parent and child grid needed for convergence for several refinement ratios and variances of Ln(T).

Practical Lessons Learned

The preceding example demonstrates five important characteristics of local grid refinement when using LGR.

1) Local refinement improves the parent-model accuracy. This is accomplished by the feedback from the child grid – the child grid can represent more accurately the hydraulics in the interior and this hydraulic information is contained in the fluxes it provides to the parent interfacing boundary.

2) Local refinement provides much of the improvement achievable by global refinement. The high resolution of the child grid is able to represent the features of interest in a similar fashion as a globally refined grid. While error is incurred at the boundary with the parent grid, often this is far away from the features of interest.

3) The greatest refinement ratio does not necessarily produce the most accurate solution. Depending on what is important to the modeling study, different refinement ratios should be used. The higher the ratio of refinement, the greater the error incurred at the boundary with the parent grid. However, higher resolutions may be required for accurate representation of features in the interior.

4) Local grid refinement does not decrease or increase the rate of convergence for homogeneous or heterogeneous models. This implies that the reduction in truncation error due to decreasing the grid size is not altered by the LGR method implemented in this report. The convergence rates are identical to those obtained by globally refining the grid.

5) The number of iterations between the parent and child grids varies depending on the heterogeneity and the grid discretization. Generally, between 10 to 20 iterations are sufficient for most problems. For situations where the parent grid can accurately represent the hydraulics in the interior, the child-grid feedback is not providing much new information and convergence will be faster.

Examples

LGR has been tested under a variety of conditions to evaluate its convergence properties and numerical accuracy. The three examples presented in this section all involve synthetic test cases and some evaluation of the error. Despite using a coupling that conserves mass between the two grids, an error is still introduced on the interface between the two grids because of the abrupt change in grid resolution. Therefore, it is important to evaluate the accuracy of this method compared to other methods of grid refinement. In this section, variably spaced grids and one-way coupled telescopic mesh refinement (TMR) are considered. Furthermore, calibration of ground-water models with locally refined grids has not been well studied. Therefore, sensitivity analysis and parameter estimation also are compared using different types of grid refinement using example 1. Lastly, computational requirements – computer memory and processing time – are typically of concern when simulating ground-water models, so these quantities are compared. These comparisons can be used to help modelers decide how to use LGR most advantageously.

Example 1: Two-Dimensional Steady State Test Case with Heterogeneity and Pumping – Forward, Sensitivity, and Inverse Simulations

The accuracy of the grid refinement technique presented in this work is compared with two other grid refinement techniques for simulating flow in the two-dimensional, heterogeneous, confined aquifer with a pumping well shown in Figure 15. The two other methods are a variably spaced grid and a traditional one-way coupled TMR method of local grid refinement. The heterogeneity pattern is based on a laboratory experiment described by Garcia (1995) and Mapa and others (1994). As shown in Figure 15, the system has constant-head boundaries on the left and right side of 10.0 m and 1.0 m, respectively, and no-flow boundaries along the top and bottom. A pumping well extracts 5.5×10^{-3} m^3/s from the system. Figure 16 shows the flow vectors and contours for this system when using the low contrast set of transmissivity values listed in Figure 15. Grid refinement is applied to increase the accuracy of the model in the vicinity of the well. Although this system is synthetic, and therefore limited in its applicability to real aquifer systems, the results are likely to provide insights regarding the typical performance that can be expected from applications of different methods of local grid refinement. The advantage of this test case is that the hydraulic-conductivity distribution provides a numerically difficult challenge for testing.

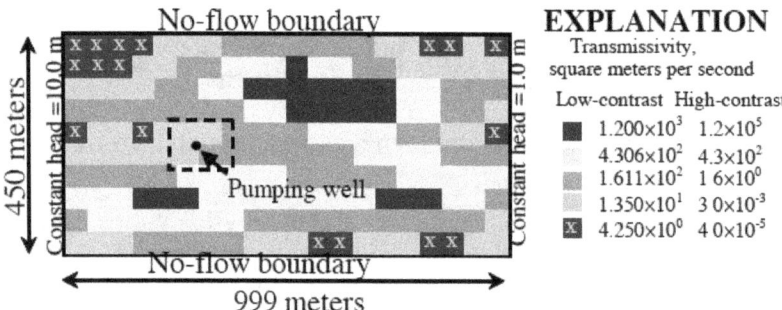

Figure 15. Heterogeneity structure and area of local refinement around the well indicated by dashed rectangle. Model results using the two sets of transmissivities are shown in Figure 16 and Figure 17, respectively.

The accuracy of the techniques are evaluated by comparing the results from a uniform fine grid ("true" solution) to (a) a variably spaced grid, (b) a TMR method of local grid refinement that does not contain a feedback but one-way couples the grids using either heads or fluxes, and (c) the iteratively coupled procedure developed for this work using both the Darcy weighted interpolation and linear interpolation of heads along the child grid boundary.

Examples

All grids were constructed such that the finite-difference cells are always fully within a single hydraulic-conductivity block. The globally fine grid has 450 rows and 972 columns, with cell dimensions of 1.028 m and a 1.0 m in the horizontal and vertical directions, respectively. The parent grid has 50 rows and 108 columns, with cell dimensions of 9.25 m and 9.0 m in the horizontal and vertical directions, respectively. The child grid has 100 rows and 154 columns with cell dimensions equal to the fine grid. The variably spaced grid has 275 rows and 380 columns; the coarsest grid spacing is never coarser than the parent grid, and the finest grid spacing is equivalent to the child grid. The traditional method of one-way coupled local grid refinement is implemented using MODTMR (Leake and Claar, 1999), and the regional and local model grids have the same spacing as the parent and child grids described above. MODTMR can be used to impose specified-head or specified-flux boundary conditions on the child model using linear interpolation from the parent model. Both options were investigated.

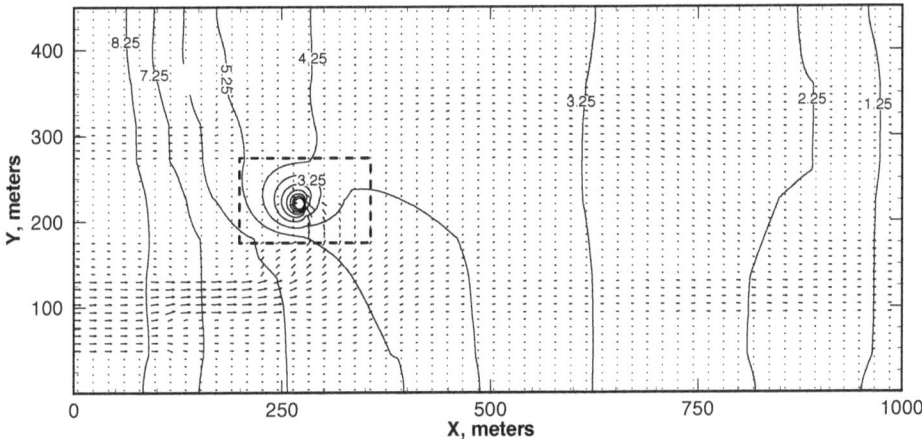

Figure 16. Flow vectors and head contours calculated using the low-contrast set of transmissivities listed in Figure 15. Area within the dashed rectangle indicates region of local refinement.

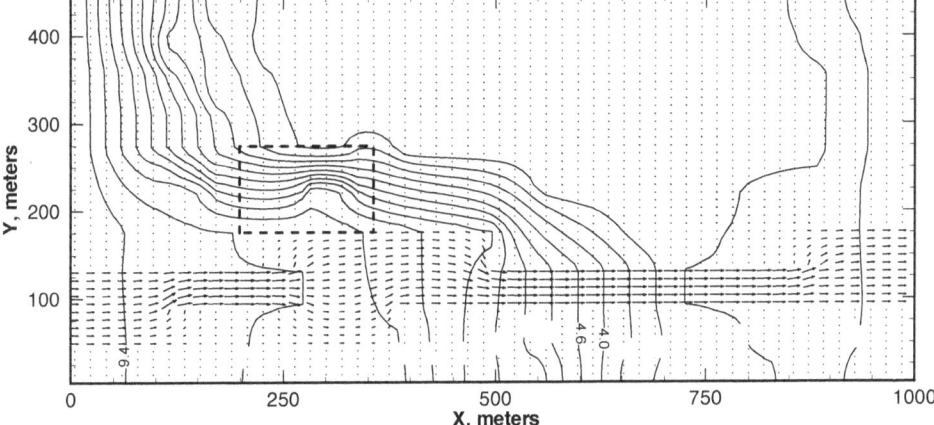

Figure 17. Flow vectors and head contours for the high-contrast set of transmissivities listed in Figure 15. Area within the dashed rectangle indicates region of local refinement.

Calculation of Heads and Flows

The results of the comparisons are shown in Table 4. All comparisons are made relative to the fine grid model, as in the previous comparisons, such that the mean head error shown in the second column is calculated as the L_1 norm of the differences between the model approximation and fine grid ("true" solution) normalized by dividing by the fine grid ("true"

31

solution). In addition to heads, the same error comparisons were made for cell-to-cell fluxes in the x and y directions; the two components were averaged to provide a single measure of cell-to-cell flux error, which is shown in the third column of Table 4 and Table 5. The CPU times for the fine grid, the variably spaced grid, and the locally refined grids were obtained using a preconditioned conjugate gradient solver, PCG2 (Hill, 1990), while the coarse parent grids were solved with a direct solver, DE4 (Harbaugh, 1995). All of the MODFLOW solvers that scale nonlinearly with grid size were evaluated, and it was found that this combination produced the fastest execution times for the computer platform indicated. Multigrid solvers such as LMG (Mehl and Hill, 2001) and GMG (Wilson and Naff, 2004) do not scale in the same way, which could cause the relative execution time advantages to be different. However, the flexibility of locally refined grids will always remain an advantage.

The results in Table 4 suggest that local grid refinement can substantially reduce the execution time, but that accuracy is diminished compared to the true solution and the variable spaced grid. For the different methods of refinement, the variably spaced grid is the most accurate, but has the longest runtime, as expected. In contrast, the traditional TMR methods have the shortest runtime, but the worst accuracy. Also, more accurate results were obtained in this case by using heads, rather than fluxes, as the interpolated boundary condition. The iteratively coupled local grid refinement method developed in this work appears to be a compromise between the results obtained with a variably spaced grid and TMR. It is faster but less accurate than the variably spaced grid, and it is slower but more accurate than TMR using either a head or flux coupling.

The results show that the Darcy-weighted interpolation is more accurate than linear interpolation. The largest differences occur for mean cell-to-cell flux errors fro the most heterogeneous test case. The small differences for results that do not include flow errors at the interface suggest that in many problems Darcy-weighted interpolation may not be important.

Effect of Heterogeneity Contrast

The degree of heterogeneity can have a substantial effect on the flow system, and consequently, on the accuracy of the local grid refinement method. These effects are investigated by using the same system as shown in Figure 15, except the magnitude of the hydraulic-conductivity field is changed such that the contrasts between materials is increased. The new values of transmissivity are shown as the high-contrast set of values in Figure 15, and the flow vectors and head contours for this system are shown in Figure 17.

Comparison of Figure 16 and Figure 17 shows that the increase in heterogeneity contrasts causes the hydraulic gradients through the system to be much steeper in some locations and change direction very rapidly, particularly in the region of local refinement.

The same methods of grid refinement were evaluated and the same solvers were used for this hydraulically more complicated system. Results are summarized in Table 5. As in the previous case, there is a trade-off between execution time and accuracy, the Darcy-weighted interpolation is more accurate than linear interpolation, and for TMR, coupling using heads rather than fluxes produces better results. Comparison of these results to those listed in Table 4 reveals that both the variably spaced grid and the iteratively coupled grid are diminished in accuracy, which is expected because of the complicated hydraulics in this system.

A surprising result is that, as measured by the mean head error, the traditional one-way coupling used by the TMR methods performs better in this more complicated scenario but have a substantially larger error in fluxes. In addition, the result of TMR coupled using heads produces a smaller mean head error than the iteratively coupled grid refinement. Additional simulations revealed that, in this case, the strict coupling and the feedback from the child to the parent actually degenerates the accuracy of the results for heads. This is because the child model is refined enough that it can accurately represent the hydraulics, so the feedback it sends the parent is representative of the complex hydraulic structure in this system. However, the parent grid is too coarse to represent the complex hydraulic structure along the interface, and thus finds a new

Examples

equilibrium such that heads and fluxes are balanced between the two grids. This is especially evident along the top (y=274.5 m) boundary of the interface between the two grids where the contour lines have a substantial change in direction over a short distance (Figure 17). This problem could be alleviated by extending the area of the child model to a region that is not so hydraulically complex or by using finer grid spacing in the parent model; in either case, the head error would decline if discretization of the parent grid was able to represent the flow field along the interface boundary. In contrast, the traditional approaches in TMR did not have this problem because there is no feedback to influence the larger grid solution. Despite having heads that are less accurate, the cell-to-cell fluxes for the iterative process are more accurate than those produced by TMR (column 3 of Table 5). This indicates that although the feedback causes an error in the heads, the gradients throughout the system are more accurate.

Table 4. Comparison of errors and computer processing time for several grid refinement methods applied to the low-contrast version of example 1.

[The system is depicted in Figure 15 with the low-contrast set of transmissivities that range nearly 2.5 orders of magnitude from 1.2×10^{3} to 4.25×10^{0} square meters per second. Computation times using a Linux workstation, Pentium II – 333MHz, 64Mb Ram as reported by Mehl and Hill (2002b); %, percent; s, second; TMR, telescopic mesh refinement; LGR, local grid refinement]

Gridding Method	Mean head error (%)	Mean cell-to-cell flux error (%)	Interior 36 percent of child model mean head error (%)	Interior 36 percent of child model mean cell-to-cell flux error (%)	Computer processing time (s)
Fine grid (truth)	0.000	0.000	0.000	0.000	716
Variably spaced	0.015	0.078	0.023	0.034	57
TMR-head[1]	0.317	4.901	0.393	2.140	3
TMR-flux[2]	3.899	17.016	6.801	7.704	4
Iterative-linear[3]	0.061	1.758	0.099	0.142	28
Iterative-Darcy[4]	0.056	1.269	0.089	0.140	28

[1]Local model uses specified-head boundary conditions derived using linear interpolation from the regional model.

[2]Local model uses specified-flux boundary conditions derived using linear interpolation from the regional model.

[3]Iterative method of LGR using a linear interpolation instead of Darcy-weighted interpolation.

[4]Iterative method of LGR using Darcy-weighted interpolation.

Table 5. Comparison of errors and computer processing time for several grid refinement methods applied to the high-contrast version of example 1.

[The system is depicted in Figure 15 with the high-contrast set of transmissivities that range nearly 10 orders of magnitude from 1.2×10^{5} to 4.0×10^{-5} square meters per second. Computation times using a Linux workstation, Pentium II – 333MHz, 64Mb Ram as reported by Mehl and Hill (2002b); %, percent; s, second; TMR, telescopic mesh refinement; LGR, local grid refinement]

Gridding Method	Mean head error (%)	Mean cell-to-cell flux error (%)	Interior 36 percent of child model mean head error (%)	Interior 36 percent of child model mean cell-to-cell flux error (%)	Computer processing time (s)
Fine grid (truth)	0.000	0.000	0.000	0.000	1,929
Variably spaced	0.043	0.190	0.031	0.116	520
TMR-head[1]	0.103	8.433	0.045	1.129	4
TMR-flux[2]	1.915	25.381	2.211	15.687	4
Iterative-linear[3]	0.171	4.143	0.084	0.903	83
Iterative-Darcy[4]	0.156	2.348	0.079	0.806	76

[1]Local model uses specified-head boundary conditions derived using linear interpolation from the regional model.

[2]Local model uses specified-flux boundary conditions derived using linear interpolation from the regional model.

[3]Iterative method of LGR using a linear interpolation instead of Darcy-weighted interpolation.

[4]Iterative method of LGR using Darcy-weighted interpolation.

Examples

Boundary Head Errors

     In this section, the accuracy of iterative and one-way coupling is tested by comparing how well the head contours mimic those of the fine-grid solution along the boundary of the refined grid. The boundary between the coarse grid and the refined grid is emphasized because it links the two grids. Figure 18 shows the head contours for this region for all methods. The contouring for the fine-grid solution was restricted to the same grid nodes as in the coarse and refined grids. Therefore, discrepancies in the contours are not artifacts of the fine-grid solution having more data points available for drawing contour lines. For example, in Figure 18a, the small offset in the 6.5-meter contour across the interface in the lower left corner results from the restriction of contoured data points and the contouring algorithm. Figure 18b, which uses heads for the one-way coupling, shows good agreement with the fine-grid solution (Figure 18a). The smooth appearance of the 6.5-meter contour across the interface of the two grids is because the heads are interpolated linearly onto the boundary of the child model and it is likely that the contouring algorithm also is based on a linear interpolation scheme.

     The larger discrepancies in the contours seen in Figure 18c demonstrate that the one-way coupling with fluxes does not provide consistent boundary conditions for heads along the child model. Lastly, the iteratively coupled method shown in Figure 18d has the best agreement with the fine-grid solution, and it also features the offset in the 6.5-meter contour at the interface of the two grids. This indicates that the iterative coupling between the grids allows the coarse-grid solution to better mimic the fine-grid solution. Such improvement in the coarse grid is not possible in the one-way coupled methods.

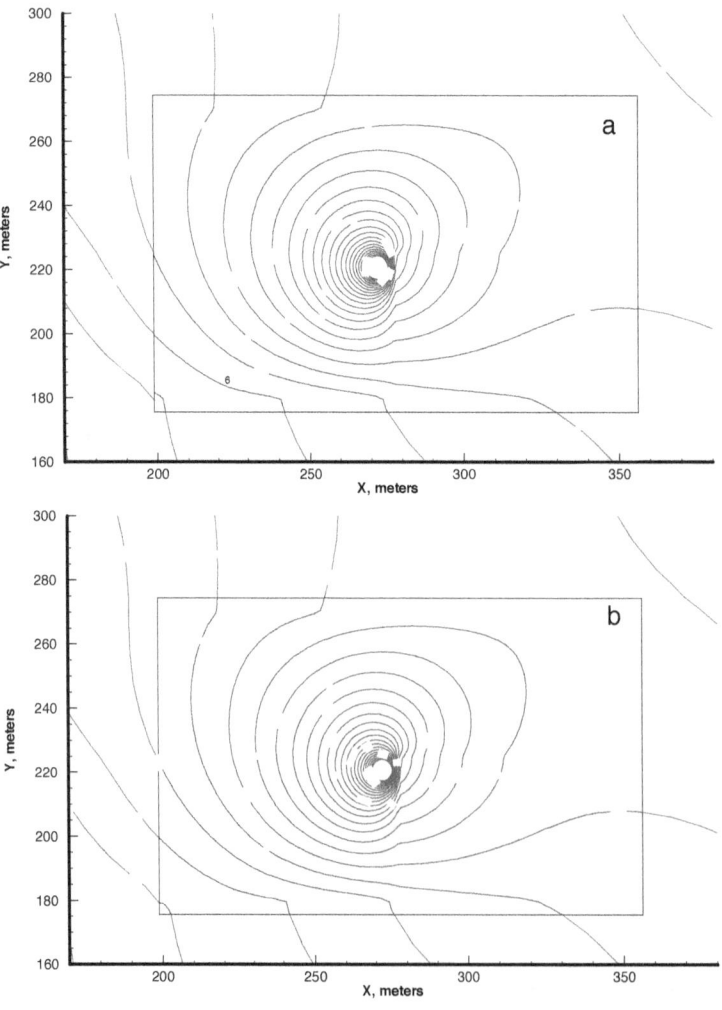

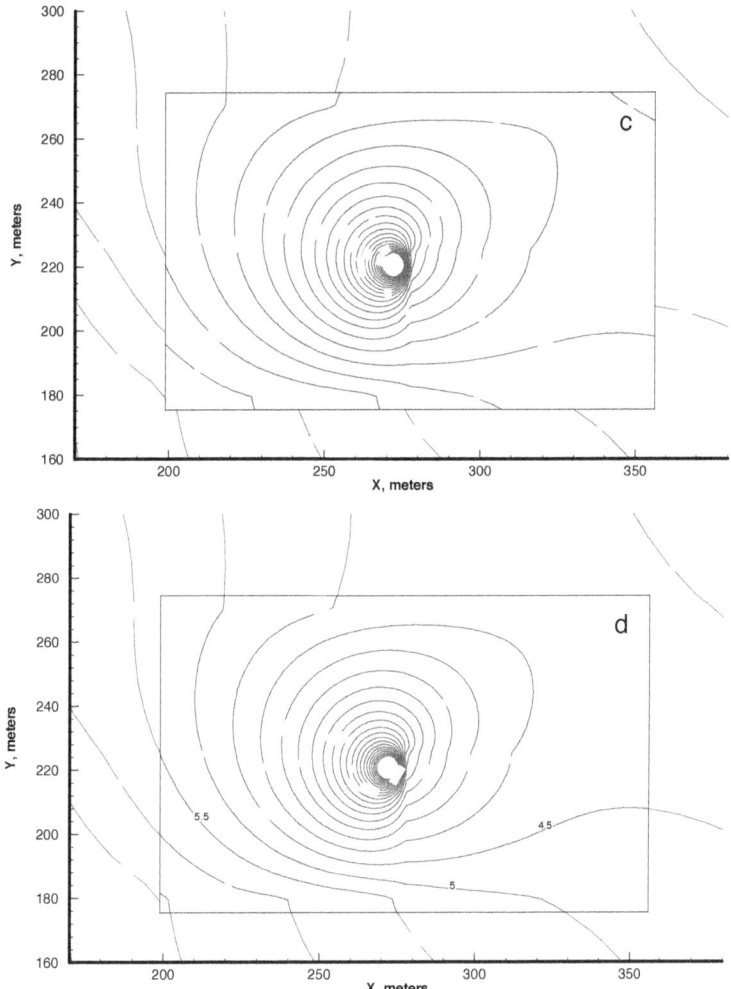

Figure 18. Head contours for (a) the fine-grid solution, a one-way coupled TMR method in which (b) heads, or (c) fluxes are used as the interpolated boundary condition, and (d) an iteratively coupled local grid refinement method. The rectangular region denotes the area of local grid refinement. The offset of the 6.5 m contour in (a) is caused by using only a subset of the fine-grid nodes for contouring.

A more formal error comparison for heads at the boundary is shown in Figure 19a. The black bars show the average of the absolute values of the percent errors for the heads along the boundary. This figure demonstrates that the coupling with fluxes has much larger errors than coupling with heads, and that the iterative coupling provides the best head solution of the methods shown.

Examples

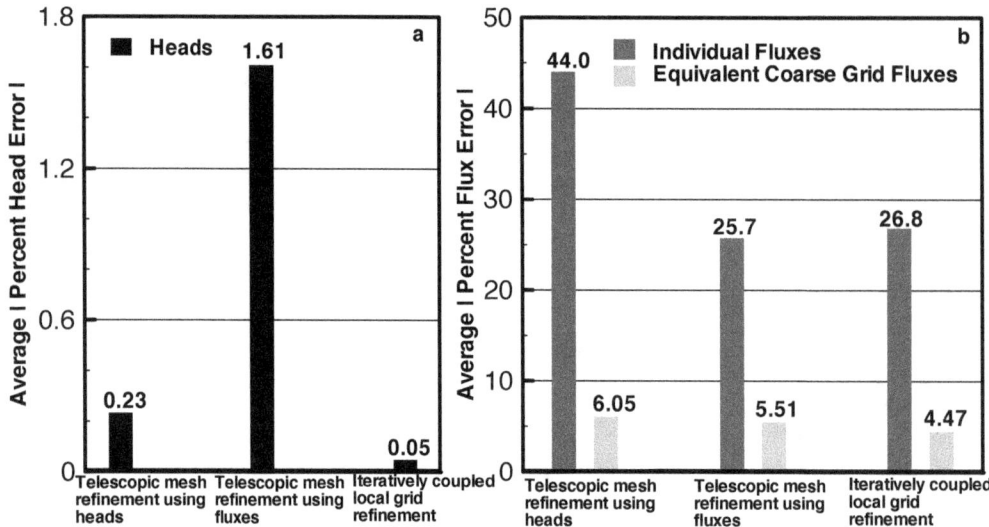

Figure 19. Errors along the boundary of the locally refined grid. Average of the absolute values of the percent errors in (a) heads, (b) individual cell-to-cell fluxes, and coarse grid equivalent fluxes.

Boundary Flux Errors

In the previous section, all comparisons were made relative to heads. It is reasonable to expect that using heads for the one-way coupling will have a better match (in terms of head) with the fine-grid solution than one-way coupling using fluxes, as was seen in the previous results. However, for many models (for example, where advective transport is of interest), accurate fluxes are more important than accurate heads. Using fluxes to couple the grids might produce models with more accurate fluxes. This section examines how well the fluxes across the boundary of the two grids match those of the fine-grid solution.

The average of the absolute values of the percent flux errors for the entire boundary is shown in Figure 19b. It is apparent that one-way coupling using fluxes more accurately reproduces flows along the boundary than coupling using heads. Even though the iterative coupling uses heads to couple to the child model, the boundary fluxes are still as accurate as those obtained using fluxes to couple the one-way model. This is not surprising because the feedback in the iterative coupling is based on fluxes, which ensures that both heads and fluxes are consistent between both grids. To provide a more thorough understanding of the boundary flux errors, Figure 20 shows the percent error in fluxes (relative to the fine grid) along the bottom boundary of the child model for all three methods. The other boundaries had similar results. A feature that is apparent in Figure 20 is the large spikes in the one-way coupling using heads. The iteratively coupled method also has these spikes, but they are less dramatic than the TMR-heads result, which indicates the feedback is able to attenuate the errors. Further examination shows a distinctive pattern in these spikes, which corresponds to the spacing between coarse grid cells.

The patterns in Figure 20 suggest that it is important to consider fluxes integrated over the width of each of the corresponding coarse grid cells to determine how well the fluxes are balanced across the interfacing coarse grid. Thus, in addition to comparing fluxes through the boundary for each individual refined grid cell, comparisons can be made based on the fluxes through the boundary of the corresponding equivalent coarse grid cells. These results (Figure 19b) show that all methods perform much better using this measure. This indicates that, while the individual fluxes along the boundary may be in error (as high as 44 percent), the characteristics of the regional flow system represented by the coarse grid are reasonably maintained. The result for the iteratively coupled method is the best, but an error of 4.47 percent is still somewhat disturbing. The next section further examines the implications of the boundary errors.

Examples

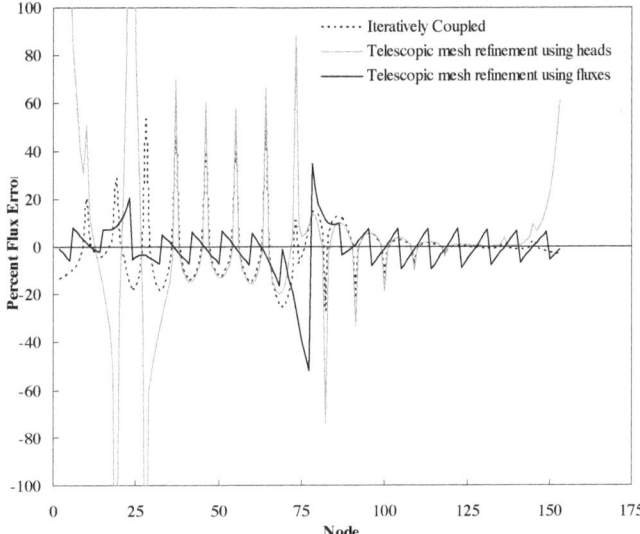

Figure 20. Percent flux errors for nodes along the bottom boundary of the embedded model.

Interior Errors

Often the accuracy within the interior of the child model is of greater importance because the locally refined region is generally of most interest. This issue was examined by comparing the errors within the interior 36 percent of the child model. The results are shown in columns 4 and 5 of Table 4 and Table 5.

For the first set of transmissivities, the results in Table 4 show that the average head error is larger in the interior for all the methods. This is because the rapid changes in gradient near the well are difficult to represent for all of the methods. In contrast, the flows are more accurate because they all must represent the same flow to the well. As expected, the variably spaced grid is the most accurate. These results also show that the iterative coupling substantially reduces internal head and flux errors relative to either of the TMR methods. Although one-way coupling with fluxes produces more accurate fluxes along the boundary (Figure 19b and Figure 20) it does not produce more accurate fluxes in the interior. Further examination of the flux errors indicated that there was a consistent bias in the flux solution, which then lead to larger errors throughout the interior of the domain.

For the larger contrasts in heterogeneity, the results in Table 5 show that generally, the ,errors are less in the interior of the child model because the influence of the well is minor and the hydraulics are more complicated near the perimeter of the refined model. These results also show the same trend – TMR-coupled using heads produces heads that have less error, but fluxes with more error than the iteratively coupled methods. This, of course, is the disadvantage of not having a rigorous coupling; there is no guarantee that both heads and fluxes are consistent between the grids.

Calculation of Sensitivities

Using the test case shown in Figure 15 and the first set of transmissivities, sensitivities are calculated for 96 head observations. UCODE (Poeter and Hill, 1998) was used to generate the sensitivities by central differences and perturbations of 5 percent. Other perturbation values were tested and it was found that, for the globally refined version of this test case, a value of 5 percent produced the closest match to MODFLOW-2000's analytically calculated sensitivities (Hill and others, 2000). Of the 96 observations, 61 are located within the area of local

Examples

refinement, and 35 are outside of this area. Inverse modeling was performed with UCODE, using the same 96 observations, to estimate the five low-contrast values of transmissivity shown in Figure 15.

Sensitivities to the 96 observations for each of the transmissivities were calculated using a variably spaced grid, TMR using head as the coupling boundary condition, TMR using flux as the coupling boundary condition, and the iterative method of LGR. The root mean square of the percent errors (RMSE) relative to the fine grid (equation 10) are shown in Table 6.

$$ \text{RMSE} = \sqrt{\frac{\sum_{i=i}^{96}\left(\frac{s'_i - s_i}{s_i} * 100\right)^2}{96}} \tag{10} $$

where,

RMSE= root mean square error.

s_i = sensitivity for the ith observation calculated using the fine grid.

s'_i = sensitivity for the ith observation calculated using one of the other discretization methods.

These results demonstrate that the variably spaced grid generally has the least error, as expected, while the TMR methods coupled using fluxes have the greatest error. These results are consistent with the accuracy in heads of Table 4. The iteratively coupled method has errors that are always less than the TMR methods, which indicates that the feedback between both grids is important for accurate sensitivity calculations.

Table 6. Comparison of root mean square percent sensitivity errors of the local grid refinement methods for the transmissivities shown in Figure 15.

[T, transmissivity (in square meters per second); %, percent; TMR, telescopic mesh refinement; LGR, local grid refinement]

Refinement method	$T = 1.2{\times}10^3$ % Error	$T = 4.306{\times}10^2$ % Error	$T = 1.611{\times}10^2$ % Error	$T = 1.35{\times}10^1$ % Error	$T = 4.25{\times}10^0$ % Error
Variably Spaced	13.60	0.1288	0.9437	0.2550	0.1321
TMR-head[1]	21.47	1.766	7.604	6.036	3.305
TMR-flux[2]	16.24	2.066	60.22	105.2	5.577
Iteratively coupled[3]	10.58	1.380	0.9457	1.630	2.921

[1]Local model uses specified-head boundary conditions derived using linear interpolation from the regional model.

[2]Local model uses specified-flux boundary conditions derived using linear interpolation from the regional model.

[3]Coupling and interpolation used by LGR

Parameter Estimation

The availability of parameter-estimation software, such as UCODE (Poeter and Hill 1998), PEST (Doherty, 2004), UCODE_2005 (Poeter and others, 2005), and OSTRICH (Matott, 2005) has made inverse methods for ground-water model calibration increasingly popular in recent years. The previous sections showed that the different methods of local grid refinement produce differences in heads, flow, and sensitivity to heads. An important question is how these differences affect model calibration. This section addresses this question by comparing inverse modeling results for the three local grid refinement schemes compared previously. The success of inverse modeling depends both on the accuracy of the sensitivities and the inverse algorithm used. The influence of forward- and central-difference sensitivities is investigated by using the more computationally frugal but less accurate forward-difference method (fwd), as well as centered (cen) differences. Three optimization algorithms are investigated using UCODE: 1) modified Gauss-Newton (G-N), modified Gauss-Newton with quasi-Newton (Q-N) updating, and the double-dogleg (DOG) trust region approach (Dennis and Schnabel, 1996), which was added to UCODE for this investigation. The trust region approach is available in UCODE_2005.

The 96 head observations used were generated, without adding noise, using the fine-grid discretization and the parameter values shown in Figure 15. The same set of starting parameter values, which were changed from the true values shown in Figure 15, were used in all simulations. The inverse modeling was evaluated on the basis of how well the 5 transmisivities returned to their true values, thus lowering the sum-of-squared residuals.

For this test case, Figure 21 shows how the sum of squared residuals is lowered as the regression proceeds and the total number of function evaluations at the final iteration, for each of the discretization methods. The more accurate central differencing is important for good convergence of the G-N method when using the variably spaced grid (Figure 21a), but has little effect in the other situations. These results also demonstrate that the double-dogleg approach can substantially reduce the number of function evaluations needed for convergence, thus reducing CPU time. The advantages of the Q-N updating were less dramatic and reliable.

The inaccurate sensitivities calculated using the TMR methods (Table 6), hindered convergence of the G-N method, and even the more accurate central differencing was unable to alleviate this problem (Figure 21b and Figure 21c). For these cases, parameter estimation sometimes converged but the resulting parameter estimates were grossly different than the true parameter values. Good convergence was achieved using either Q-N updating or the double-dogleg strategy in the case of coupling with heads.

The sensitivities calculated by UCODE using the iteratively coupled method of LGR were accurate enough that good convergence was achieved for all scenarios (Figure 21d). Indeed, convergence equaled or exceeded that of the variably spaced grid.

Considerations for Inverse Modeling with Locally Refined Grids

These results demonstrate that different methods of local grid refinement can have a substantial effect on parameter-sensitivity calculations. The results also show that the accuracy of the sensitivity calculations influence the regression, and that some of the inaccuracy can be overcome by using more sophisticated search techniques, as shown most notably by the double-dogleg trust region results. Similar difficulties in estimating parameters with error-prone models can be expected if the parameters are estimated by trial and error; use of regression just makes the consequences more obvious.

For the case presented, the variably spaced grid method of local grid refinement is most accurate in terms of sensitivity calculations, while the TMR methods were least accurate. This is because the TMR methods do not provide a way for changes to parameter values of the child grid to be fully accounted for in the regression if there are observations in the parent model. For example, if changing hydraulic conductivities in the child grid increases the bulk flow rate through that grid, the extra flow should be coming from the regional flow system represented by the parent grid. However, TMR methods cannot account for this because the heads and fluxes of the regional model are independent of the child grid. If different parameters are defined within the local (child) model that do not have equivalents in the regional (parent) model, lack of a feedback means that there is no information available to the regional model regarding these parameters. In this regard, example 1 may be viewed as a "best-case scenario" because all of the parameters of the local model also were represented in the regional model.

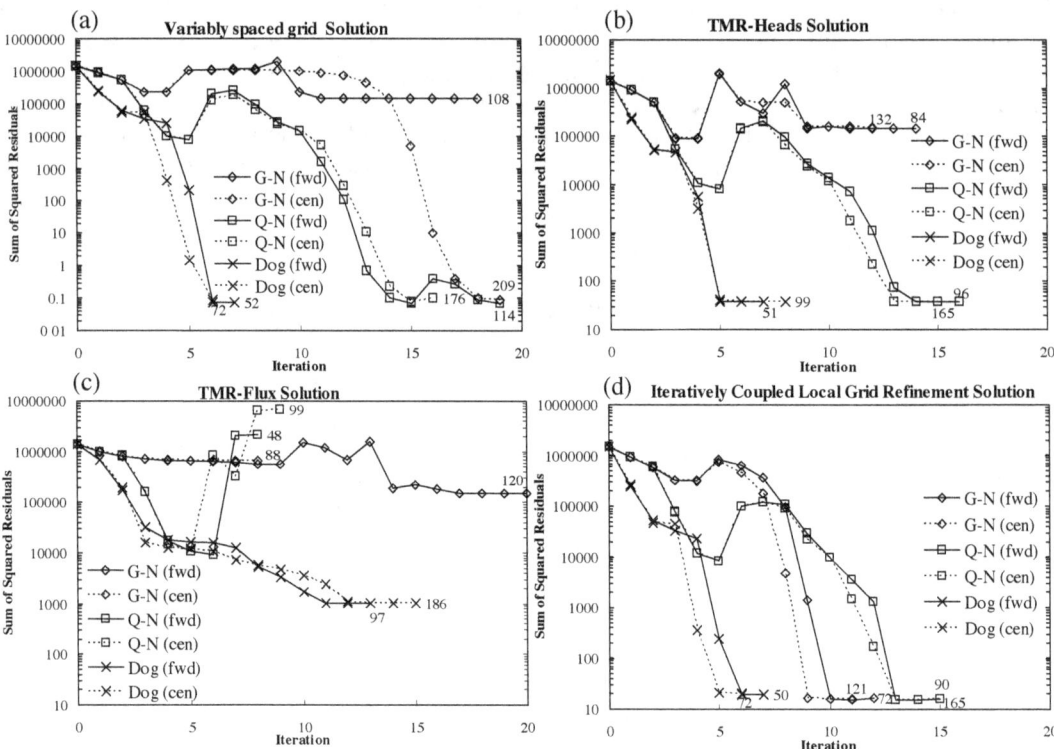

Figure 21. Regression results using the Gauss-Newton (G-N), quasi-Newton (Q-N), double-dogleg (DOG) methods with forward (fwd) and central (cen) differencing for simulations using a (a) variably spaced grid, (b) telescopic mesh refinement (TMR) coupled using heads, (c) TMR coupled using fluxes, and (d) an iteratively coupled local grid refinement. Total number of function evaluations is indicated at the final iteration.

For the case presented, iterative coupling always was more accurate than the TMR methods of the same grid discretization and approached the accuracy of the variably spaced grid. The iterative feedback of LGR means the effects of changes within the local grid are accounted for in the regional grid, whether or not the parameter appears in both the regional and local models. This decreases some of the burden on the modeler in that parameter sensitivities are consistent for both grids regardless of their presence (or lack there of) in each model so that the modeler does not have to devise a method to update the parent grid to account for parameter changes in the child grid.

Example 2: Two-Dimensional Transient Test Case with Homogeneity and Pumping – Forward Simulations

A modification of the homogeneous test case that was used for the convergence analysis (shown in Figure 9 and Figure 10) is used to demonstrate the performance of LGR for simulating transient flow. For this demonstration, the specified-head boundaries are fixed at 10.0 m on the left and right sides. An initial head distribution of 10.0 m is used throughout the aquifer before pumping begins at a rate of 270 m^3/hr. Transmissivity and specific storage are set to 2.5 m^2/hr and 2.0×10^{-5} m^{-1}, respectively. Six stress periods are used with lengths of 1, 1, 8, 10, 80, and 100 hours; five time steps are in each stress period.

The same set of locations shown in Figure 10, including the node of the pumping well, are used to compare heads in the child part of the grid using both locally and globally refined grids. Because of the modifications to the boundary conditions, the transmissivity, and the pumping rate, the error at the pumping well does not dominate and is therefore included in the analysis. The globally refined grid has the same grid spacing as the child grid over the entire domain. Head errors are measured as the average L_1 norm of the percent difference between results simulated using the globally and locally refined models (sum of the absolute values of the percent differences divided by 361); comparisons are made at five times: 1, 2, 10, 20, 100, and 200 hours.

Here, a one-way coupled method is simulated by using LGR and setting the maximum number of iterations between the parent and child grids to one. Figure 22 shows that for both the iterative and one-way coupling, the errors of the child grid peak around 10 hours while the smallest errors occur early and late in the simulation when the system is still near its initial and steady-state conditions, respectively. Figure 22 shows that, generally, the error is reduced by iterating and this error is reduced as the grids are refined. For example, for the 15×15 grid (Figure 22a), errors are between 0.7-10 percent, while the head errors for the 135×135 grid (Figure 22c) are between 0.006 and 0.2 percent. In contrast, when one-way coupling is used, the errors are not substantially reduced as the grids are refined; they remain between 0.7 and 15 percent. This is because the one-way coupled method does not feedback improvements to the parent grid; this is elaborated on using example 3.

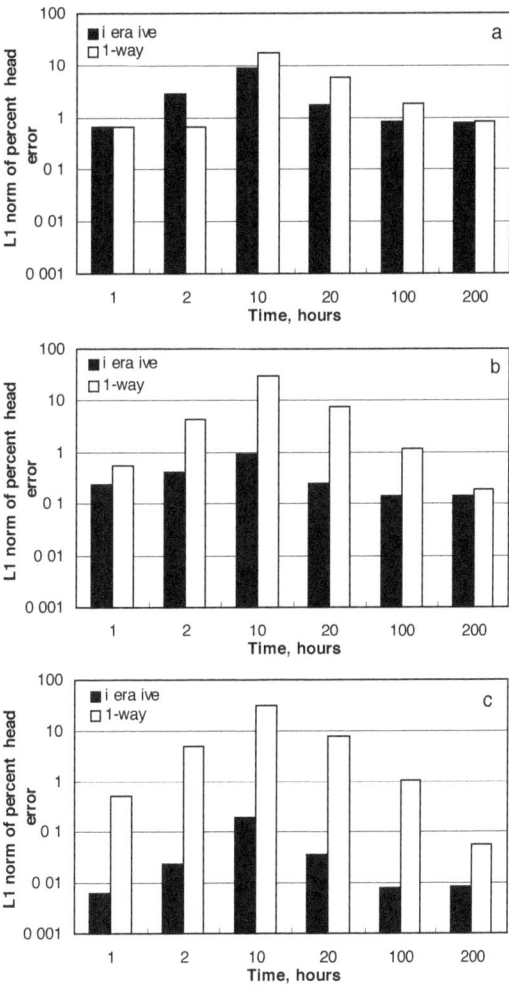

Figure 22. Average L_1 norm of percent head errors (sum of the absolute values divided by 361) between globally and locally refined grids at five times using iteratively and one-way coupled grids. The 361 head comparison locations are in the child grid and shown in Figure 10. Child grids are a 3:1 refinement of the parent. Parent grid resolutions are (a) 15×15 cells, (b) 45×45 cells, and (c) 135×135 cells.

Example 3: Three-Dimensional Steady State Test Case with Homogeneity and Stream-Aquifer Interactions – Forward Simulations

This test case demonstrates the performance of LGR in a three-dimensional, unconfined aquifer. This test case also is used to examine the change in the errors as the LGR iterations progress.

The hypothetical ground-water model used in this analysis is shown in Figure 23. The meandering stream has a total length of 3,409 m and has a linear drop in stage along the length of the river from the inlet at 50.0 m to the outlet at 45.0 m. This results in a gradient along the river of 0.00147. The width, thickness of the streambed, and the streambed hydraulic conductivity are constant throughout the length at 1.0 m, 0.5 m, and 1.0 m/day, respectively. The land-surface elevation of the model domain follows a linear profile from 50 m at the left boundary and drops to 45 m at the right boundary. The bottom elevation also follows this linear profile such that the model has a uniform thickness of 50 m throughout the domain. The specified-head boundaries at both ends provide a background gradient equal to the slope of the top and bottom of the model (0.00347). The aquifer is homogeneous and isotropic with a hydraulic conductivity of 1.0 m/day. The system is unconfined, which causes nonlinearity in the flow because the saturated thickness depends on the value of head, which is not known beforehand.

Three sets of parent-child grids of different resolutions were applied to the area shown in Figure 23. The coarsest parent grid used is 15×15×3 and the child model is a 3:1 ratio of refinement, which results in a 19×22×5 child grid, as shown in Figure 23. The refinement also extends in the vertical direction from the top of the model down to the middle of the aquifer, which accounts for the top 1 ½ layers of the three-layer parent model. Thus, in the refined region, a single parent cell is replaced by 27 child cells. 45×45×3 and 135×135×3 parent grids also were used with child grids of a 3:1 refinement ratio, resulting in child grids of 55×64×5 and 163×190×5. Using each of these grids, the hypothetical aquifer was simulated and the errors examined.

Head and Flux Convergence and Analysis of Errors

The goal of local grid refinement is to approach the accuracy of a globally refined grid. Thus, for this analysis, each comparison is made to results obtained from a globally refined model – a model with grid spacing equivalent to the child grid over the entire domain. Both heads and fluxes are examined and results are presented in Figure 24 and Figure 25. Heads are compared at 2,090 locations that correspond to all of the nodes of the coarsest child grid. Head errors are measured as the average L_1 norm of the percent difference between results simulated using the globally and locally refined models (sum of the absolute values of the percent differences divided by 2,090). This error measure is plotted for heads in Figure 24 and indicates how the solution in the interior as well as the boundary changes as the iteration proceeds.

Examples

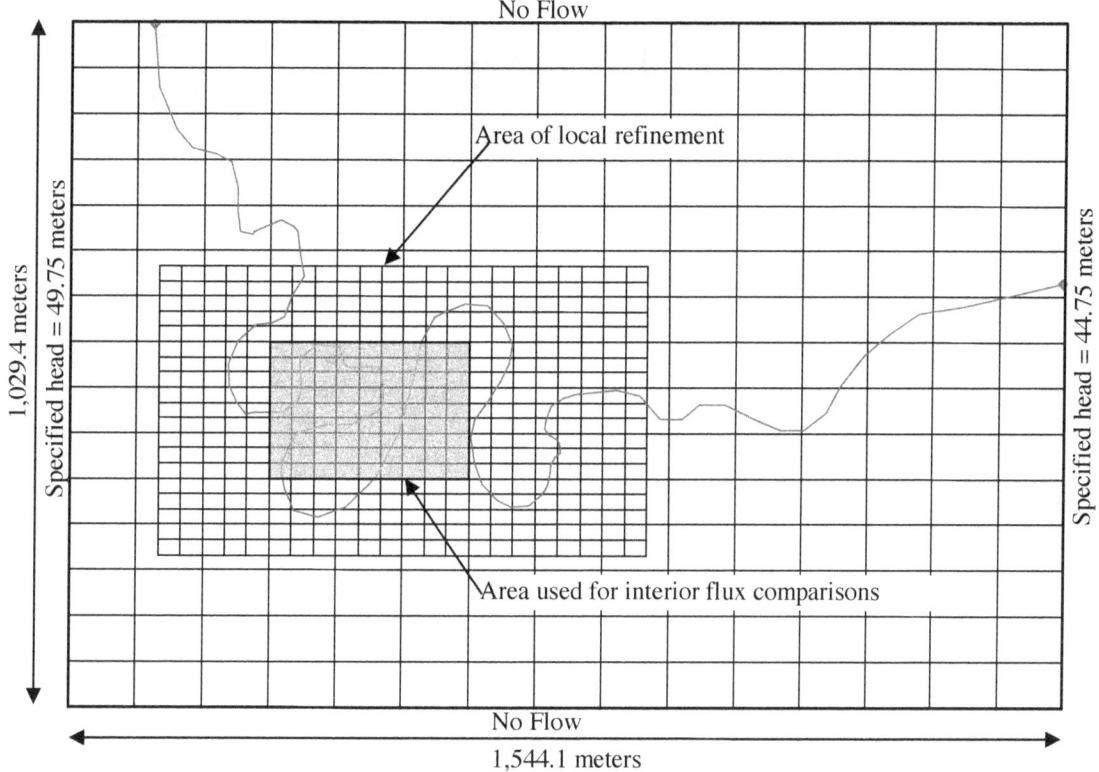

Figure 23. Plan view of a three-dimensional aquifer system used to test the local grid refinement method. A 15×15 horizontal grid discretization is shown for the parent grid and the locally refined grid (19×22) spacing is equivalent to a 45×45 discretization over the whole domain. The shaded area is used to evaluate the accuracy of the interior fluxes within the child grid.

Fluxes are compared along the parent-child boundary in Figure 24. The fluxes along the parent-child boundary are the same fluxes used as a feedback from the child to the parent model. Comparing these fluxes indicates how the error in the coupling between the grids changes as the iteration proceeds. Because a few individual fluxes that are small in magnitude dominate a percent difference comparison, the difference between the globally and locally refined boundary fluxes is instead compared using the net flux in and out of the refined region. The absolute value of the percent errors of the net flux in and net flux out were then averaged, as shown in equation 11.

$$\text{Average Flux Error (percent)} = \frac{1}{2} \times \left[\left| \frac{q_l^i - q_g^i}{q_g^i} \right| + \left| \frac{q_l^o - q_g^o}{q_g^o} \right| \right] \times 100 \qquad (11)$$

where,

q is flux,

superscripts "i" and "o" correspond to inflow and outflow, respectively, and subscripts "l" and "g" correspond to locally and globally refined grids, respectively.

Figure 24 shows how the head errors throughout the child domain and the flux errors along the parent-child interface change with each iteration of the parent-child coupling. In this example, the starting hydraulic heads of the child model are equal to the heads from a parent-grid resolution model that includes the entire domain. The errors from the coarse-grid solution are shown at iteration zero. The errors from iteration one are the same as those produced by models that are one-way coupled using heads.

In all cases, the errors oscillate in early iterations before they diminish. This oscillation probably results from the iterative coupling between heads and fluxes – it tends to update the heads and fluxes in the correct direction, but overshoots. The relaxation applied between successive iterations keeps the errors from growing and the oscillations eventually diminish. The solution stabilizes after 10 iterations for the cases presented and further iterations do not improve the solution. This number of iterations is reasonably consistent with what was found for the homogenous test cases analyzed in Figure 14. At this point, the parent and child grids are in equilibrium at the interfacing boundary – interpolated heads from the parent grid to the boundaries of the child grid result in a child grid simulation that produces fluxes across the interfacing boundary that are consistent with the parent head solution.

For this test case, the head errors are small for all iterations, but the flux errors are large enough to be of practical concern. Errors in flux are reduced to 2 to 5 percent with iteration. As expected, comparison of Figure 24a, Figure 24b, and Figure 24c show that the magnitude of the errors in both heads and fluxes generally decreases as the resolution increases. A noticeable exception occurs for the flux errors at iteration one for all the grids. The flux error at iteration one is actually larger in Figure 24c than Figure 24b, which indicates that increased resolution does not alleviate this problem. This suggests that one-way coupled approaches can produce poor results unpredictably because the results used are from the oscillatory stages of the coupling where the grids are not in equilibrium.

Table 7 shows the ratio of error reduction of the locally refined grid to the coarse grid for the head and flux solutions. The errors are those shown in Figure 24, which are calculated relative to the globally refined grid solution. In most cases, the local refinement reduces the errors compared to the coarse grid. The exception occurs for the one-way coupled method where the flux errors along the boundary are actually larger than if no refinement were used. In the one-way coupled approach heads were used to couple the grids. In contrast, if fluxes are used to couple the grids, the flux errors along the boundary are lower, but the head errors are worse. Furthermore, there is no guarantee that improved fluxes at the boundary produce a better flux solution in the interior of the model (Figure 19b and Table 4).

In the iterative procedure used in LGR, the errors in head and flux influence each other until a solution is achieved where the two grids are in equilibrium. This raises the question: how do the parent and child grids interact to change the head solution so that equilibrium is achieved? To address this question, the iterations were started with the globally refined head solution and the resulting head and flux errors were tracked. These results are shown as dashed lines in Figure 24.

Examples

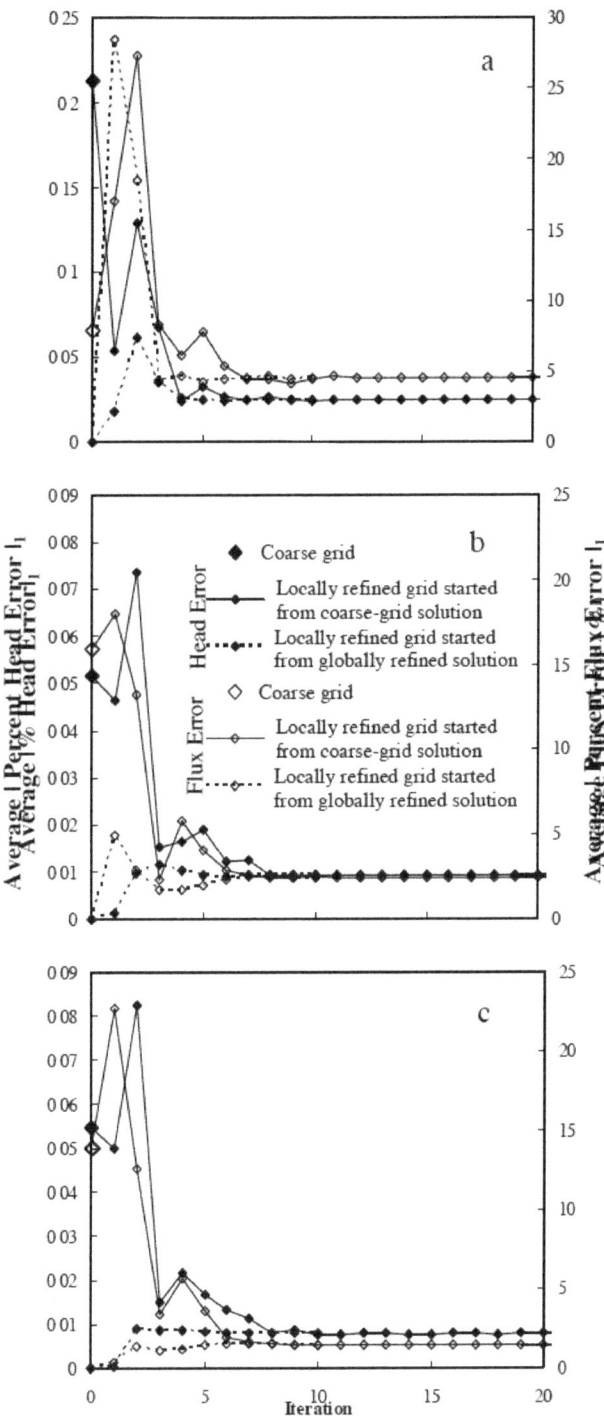

Figure 24. Head and flux errors along the parent-child bounary in relation to number of iterations of the local grid refinement procedure for parent grid resolutions of (a) 15×15×3, (b) 45×45×3, and (c) 135×135×3. All child grids are a 3:1 refinement of the parent. All errors are relative to the globally refined solution. Dashed lines represent results from locally refined grids that were started with the head solution from the globally refined grid. |·|₁ represents L_1 norms (absolute values) for 2,090 heads throughout the refined region. Flux errors are average error of the net inflow and net outflow. Iteration one is equivalent to a one-way coupled with heads method.

Examples

Table 7. Error ratios of the coarse grid, one-way coupled, and iteratively coupled solutions.

[Values less than one indicate that using the first method improved the solution relative to the second method. Values greater than one indicate a worse solution and are shaded. Errors are calculated relative to the globally refined grid.]

Parent Grid[1]	rowsxcolumnsxlayers 15×15×3		rowsxcolumnsxlayers 45×45×3		rowsxcolumnsxlayers 135×135×3	
Methods Compared[2]	Head Error	Flux Error	Head Error	Flux Error	Head Error	Flux Error
One-way/ Coarse	0.254	2.17	0.897	1.13	0.917	1.63
Iterative/ Coarse	0.117	0.585	0.181	0.153	0.143	0.108
Iterative/ One-way	0.460	0.270	0.202	0.136	0.157	0.0661

[1]Child grids are a three times refinement of the parent grid and cover the area shown in Figure 23.

[2]The error ratios are calculated as the error for the first method divided by the error for the second method. The errors are shown in Figure 24 started from the coarse-grid solution. The "Coarse" method errors are at iteration zero, the "One-way" method errors are at iteration one, and the "Iterative" method errors are at iteration twenty.

The interacting of parent and child grids shown in Figure 24 for the simulations started from the globally refined solution reveal a number of features that are likely to be characteristic of all locally refined grids. By definition, the globally refined grid head and flux solution error is zero at iteration zero. The head error at iteration one is due solely to the error introduced from the two-step cage-shell interpolation along the boundary – the globally refined heads are imposed at the shared nodes, the two-step cage-shell interpolation procedure is applied to obtain the child specified-head boundary conditions, and the child simulation is completed. If the interpolation were perfect, the head and, therefore, the flux error at iteration one would be zero. Investigating the error further, the interpolation produces fluxes between child nodes of the boundary that are consistent with the parent-grid fluxes. The parent fluxes are constant between adjacent shared nodes (parent cells), and this forces the head interpolation to have the structure of the Darcy-weighted interpolation shown in Figure 6 – piecewise linear between shared nodes. The child grid has more than one cell between shared nodes, and potentially could represent curvature in the head profile between the shared nodes, but such curvature results from a change in flux along the length between the shared nodes which cannot be represented in the parent grid. Therefore, the child-grid boundary cells cannot represent such curvature given the premise of the cage-shell approach. As the parent grid resolution is increased, the distance between the shared nodes decreases, and the piecewise linear approximation becomes more accurate. This is reflected in the smaller errors of the globally refined grid head solutions at iteration one relative to their final values.

The error in head interpolation is propagated into the interior and causes the fluxes to deviate from their true values, which can be seen at iteration one of the globally refined grid fluxes line of Figure 24. This error in flux gets fed back and affects the parent-grid flux boundary condition and causes further errors in head at the subsequent iteration. The cycle continues until the same equilibrium is reached as was achieved when starting with the coarse-grid solution – the self correcting nature of the coupling finds the same equilibrium solution despite starting from a different initial guess. The results above show that even if the true heads or fluxes were known, the other quantity would be in error.

Table 7 shows that for the one-way coupled method, an accurate head solution may not imply an accurate flux solution. This underscores the importance of using methods that balance both heads and fluxes when applying local grid refinement.

The shift from the globally refined solution to one that satisfies both heads and fluxes in both grids is not unique to the iteratively coupled method of LGR. The directly coupled techniques, for example, Wasserman (1987), Ewing and others (1991), Edwards (1999), Schaars and Kamps (2001) and Haefner and Boy (2003), produce solutions that satisfy the modified equations at the interface, and these result in solutions that are different from the globally refined solution.

In a one-way coupled approach, the shifting away from the true solution would not occur (although some error could be introduced from the interpolation), but of course, one never has the luxury of starting with the true solution and instead typically starts with the coarse grid solution. Table 7 demonstrates that the error reduction achieved by iterating compared to a one-way coupled approach improves as the refinement increases. This is probably due to the improved accuracy of the parent grid as the grids are refined, which causes the interpolation procedure to be more accurate and provides better consistency between the parent and child grids. In other words, as the grids are refined, the parent grid is better able to represent the hydraulics in the feedback from the child grid. In contrast, the one-way coupled approach shows much less improvement over the coarse grid as the grids are refined. Because the additional accuracy gained by the refined region is not fed back to allow further improvement, the accuracy is limited by the error in the head boundary conditions from the parent being propagated and diffused through the child grid (see Error Propagation in LGR in Appendix 3).

Interior Errors

When starting with the parent-grid solution, Figure 24 shows that, at iteration one – which is equivalent to a one-way coupled approach – the head error has decreased, but the flux error along the boundary has increased relative to iteration zero. Because the coupling is based on a head interpolation, a better match to heads than fluxes is expected, but it is surprising that the flux error is worse than the coarse-grid solution. Often, the phenomena of interest are within the child grid, so an important question is, "could the one-way coupled approach produce an interior solution that is worse than having no refinement at all?" To address this question, the shaded area in the interior of Figure 23 was used to perform a flow budget analysis into the river, the lateral (side) boundaries, and the bottom boundary using ZONEBUDGET (Harbaugh, 1990). These results are shown in Figure 25. The bottom boundary of the shaded area corresponds to the bottom of the first layer of the three-layer parent models, and is above the bottom of the child model which, in the shared node method of LGR, extends to the node of the second layer of the parent model.

Figure 25 shows that the one-way coupled approach generally produces more accurate fluxes in the interior than the parent grid solution. An exception occurs for the 15×15 grid at the bottom boundary of the interior region. The reason for the poor results on the bottom boundary in all the locally refined models is because it is closest, in terms of the number of cells, to the parent-child interface. The boundary flux errors shown in Figure 25 are attenuated when moving away from the boundary. As the discretization increases, the number of cells from the interior to the boundary increases which insulates the interior cells from the boundary errors. In this example, the parent and child models have increasing horizontal resolution which insulates from the side boundaries; however, the parent and child always have three and five layers, respectively. Thus, the number of cells insulating from the bottom boundary stays the same. The iteratively coupled approach shows less susceptibility to the bottom-boundary errors, particularly as the grid is refined.

For head error, the maximum error always occurs at the boundary and is propagated into the interior with the same properties of the governing ground-water flow equation (a diffusion equation) minus sink/source terms (see Appendix 3). Therefore, in general, locations further from the interfacing boundary have greater opportunity to decrease the head error.

In many ground-water models, the number of rows and columns is much greater than the number of layers. This restricts the location of the refinement in the vertical direction and limits the possibilities of insulating from errors that arise from the coupling with the bottom boundary. Although not done here, many local grid refinement studies use a hierarchy of refinement (parent, child, grandchild, and so forth). The quality of the interior solution is important because it will be used to provide boundary conditions to another local model. It is in these two cases

Examples

that the consistency and accuracy of the boundary conditions of the iteratively coupled shared-node method are particularly advantageous.

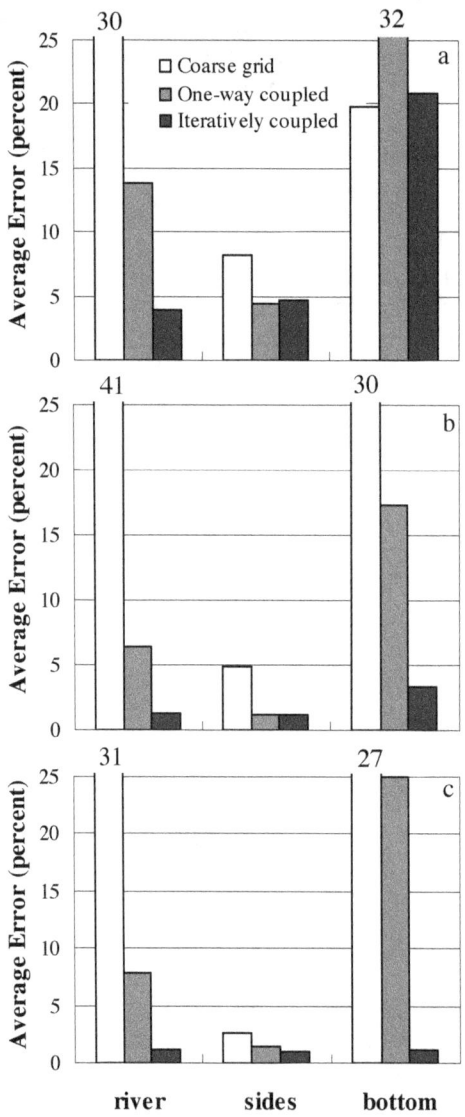

Figure 25. Flux errors evaluated for the river, sides, and bottom of the volume in the interior of the locally refined grid (shaded area shown in Figure 23 extends 1/3 of system depth). Parent grid resolutions are (a) 15×15×3, (b) 45×45×3, and (c) 135×135×3. All child grids are 3:1 refinement of the parent. Errors are relative to the globally refined grid solution.

Computational and Accuracy Comparisons

In addition to accuracy, CPU time and memory requirements are typically of concern for ground-water models. Figure 26 shows this tradeoff for the finest grids used in example 3. This figure demonstrates that a substantial savings in CPU time and memory can be achieved by using local grid refinement instead of globally refining the grids. The iteratively coupled method requires that both the parent and child grids need to be solved many times, therefore its CPU time requirements are larger than the one-way coupled method which only solves each grid once. In this example, 11 iterations were used and the CPU time required is about 7 times greater than the one-way coupled method. The reason why it is not 11 times greater than the one-way coupled method is because the previous solution is used as the starting guess for subsequent solutions and the solvers tend to converge more quickly as the iterations proceed. Furthermore, the iterative coupling stores the information for both grids in memory; therefore, its memory requirements are slightly larger than one-way coupling which only store information for one grid at a time (for one-way coupling, Figure 26a only lists the memory required for the child grid, which has more nodes than the parent grid).

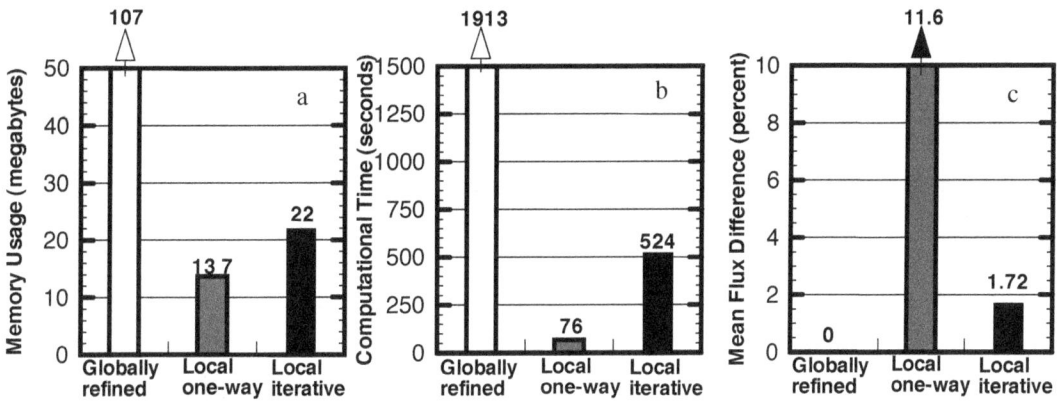

Figure 26. Comparison of (a) memory requirement, (b) computer processing time, and (c) accuracy of stream fluxes using globally and locally refined grids with one-way and iterative coupling for example 3. Parent grid is 135×135×3, child grid is 163×190×5, and the globally refined grid is 405×405×9.

Literature Cited

Anderman, E.R. and Hill, M.C., 2001, MODFLOW-2000, the U.S. Geological Survey modular ground-water model – Documentation of the ADVective-Transport Observation (ADV2) Package, Version 2: U.S. Geological Survey Open-File Report 01-54, 69 p. (Available at http://water.usgs.gov/nrp/gwsoftware/modflow2000/modflow2000.html)

Anderman, E.R., Kipp, K.L., Hill, M.C., Valstar, J., and Neupauer, R.M., 2002, MODFLOW-2000, the U.S. Geological Survey modular ground-water model – Documentation of the Model-Layer Variable-Direction Horizontal Anisotropy (LVDA) capability of the Hydrogeologic-Unit Flow (HUF) Package: U.S. Geological Survey Open-File Report 02-409, 61 p. (Available at http://water.usgs.gov/nrp/gwsoftware/modflow2000/modflow2000.html)

Chan, Y.K., Mullineux, N., and Reed, J.R., 1976, Analytical Solutions for Drawdowns in Rectangular Artesian Aquifers: Journal of Hydrology, v. 31, p. 151-160.

Davison, R.M. and Lerner, D.N., 2000, Evaluating Natural Attenuation of Groundwater Pollution from a Coal-Carbonisation Plant: Developing a Local-Scale Model using MODFLOW, MODTMR and MT3D: Water and Environmental Management Journal, v. 14, p. 419-426.

de Marsily, G., 1986, Quantitative Hydrogeology: Orlando, Academic Press, Inc, 440 p.

Dennis, J.E. and Schnabel, R.B., 1996, Numerical Methods for Unconstrained Optimization and Nonlinear Equations: Philadelphia, Society for Industrial and Applied Mathematics, 378 p.

Doherty, J., 2004, PEST – Model-independent parameter estimation: Australia, Watermark Numerical Computing. (Available at http://www.sspa.com/PEST/index.htm).

Edwards, M.G., 1999, A high-resolution method coupled with local grid refinement for three-dimensional aquifer remediation: In Situ 1999, v. 23, no. 4, p. 333-377.

Ewing, R.E., Lazarov, R.D., and Vassilevski P.S., 1991, Local refinement techniques for elliptic problems on cell-centered grids I. Error analysis: Mathematics of Computation, v. 56, no. 194, p. 437-461.

Forsyth, P.A. and Sammon, P.H., 1988, Quadratic convergence for cell-centered grids: Applied Numerical Mathematics, v. 4, p. 377-394.

Funaro, D., Quarteroni, A., and Zanolli, P., 1988, An iterative procedure with interface relaxation for domain decomposition methods: Siam Journal on Numerical Analysis, v. 25, no. 2, p. 1213-1236.

Garcia, J.E., 1995, An experimental investigation of upscaling of water flow and solute transport in saturated porous media: MS Thesis, University of Colorado, Boulder, 135 p.

Haefner, F., and Boy, S., 2003, Fast Transport Simulation with an Adaptive Grid Refinement: Ground Water, v. 41, no. 2, p. 273-279.

Harbaugh, A.W., 1990, A Computer Program for Calculating Subregional Water Budgets Using Results from the U.S. Geological Survey Modular Three-dimensional Finite-difference Ground-water Flow Model: U.S. Geological Survey Open-File Report 90-392, 46 p. (Available at http://water.usgs.gov/nrp/gwsoftware/zonebud2/zonebudget2.html)

Harbaugh, A.W., 1995, Direct Solution Package Based on Alternating Diagonal Ordering for the U.S. Geological Survey Modular Finite-Difference Ground-Water Flow Model: U.S. Geological Survey Open-File Report 95-288, 46 p. (Available at http://water.usgs.gov/nrp/gwsoftware/modflow2000/modflow2000.html)

Harbaugh, A.W., 2005, MODFLOW-2005, the U.S. Geological Survey modular ground-water model – The Ground-Water Flow Process: U.S. Geological Survey Techniques and Methods 6-A16, 9 Ch.

Harbaugh, A.W., Banta, E.R., Hill, M.C., and McDonald, M.G., 2000, MODFLOW-2000, the U.S. Geological Survey modular ground-water model – User guide to modularization concepts and the ground-water flow process: U.S. Geological Survey Open-File Report 00-92, 121 p.

Hill, M.C., 1990, Preconditioned Conjugate-Gradient 2 (PCG2), A computer program for solving ground-water flow equations: U.S. Geological Survey Water-Resources Investigations Report 90-4048, 43 p. (Available at http://water.usgs.gov/nrp/gwsoftware/modflow2000/modflow2000.html)

Hill, M.C., Banta, E.R., Harbaugh, A.W., and Anderman, E.R., 2000, MODFLOW-2000, The U.S. Geological Survey modular ground-water model – User guide to the observation, sensitivity, and parameter-estimation processes and three post-processing programs: U.S. Geological Survey Open-File Report 00-184, 209 p. (Available at http://water.usgs.gov/nrp/gwsoftware/modflow2000/modflow2000.html)

Hunt, R.J., Steuer, J.J., Mansor, M.T.C., and Bullen, T.D., 2001, Delineating a Recharge Area for a Spring Using Numerical Modeling, Monte Carlo Techniques, and Geochemical Investigation: Ground Water, v. 39, no. 5, p. 702-712.

Leake, S.A., Lawson, P.W., Lilly, M.R., and Claar, D.V., 1998, Assignment of Boundary Conditions in Embedded Ground Water Flow Models: Ground Water, v. 36, no. 4, p. 621-625.

Leake, S.A. and Claar, D.V., 1999, Procedure and computer programs for telescopic mesh refinement using MODFLOW: U.S. Geological Survey Open-File Report 99-238, 53 p. (Available at http://az.water.usgs.gov/tmrprogs/tmr.html)

Mapa, R., Illangasekare, T.H., and Garcia, J.E., 1994, Upscaling of Water Flow and Solute Transport in Saturated Porous Media: Theory, Computation and Experiments: Progress report submitted to U.S. Army Waterways Experiment Station, 184 p.

Matott, L.S., 2005, OSTRICH, An optimization software tool, documentation and user's guide, Version 1.6: State University of New York at Buffalo, 114p. Accessed January 31, 2006 at http://www.groundwater.buffalo.edu/software/Ostrich/OstrichMain.html

McDonald, M.G., and Harbaugh, A.W., 1988, A modular three-dimensional finite-difference ground-water flow model: U.S. Geological Survey Techniques of Water-Resources Investigations, book 6, chap. A1, 548 p. (Available at http://water.usgs.gov/pubs/twri/twri6a1/)

Mehl, S., 2003, Development and evaluation of local grid refinement methods for forward and inverse groundwater models: PhD thesis, University of Colorado, Boulder, 160 p.

Mehl, S., and Hill, M.C., 2001, MODFLOW-2000, The U.S. Geological Survey modular ground-water model – User guide to the Link-AMG (LMG) package for solving matrix equations using an algebraic multigrid solver: U.S. Geological Survey Open-File Report 01-177, 33 p. (Available at http://water.usgs.gov/nrp/gwsoftware/modflow2000/modflow2000.html)

Mehl, S. and Hill, M.C., 2002a, Evaluation of a local grid refinement method for steady-state block-centered finite-difference groundwater models. Computational Methods in Water Resources XIV, June, 23-28, 2002, Delft, Netherlands, Developments in Water Science 47, Elsevier, Volume 1, 367-374.

Mehl, S. and Hill, M.C., 2002b, Development and evaluation of a local grid refinement method for block-centered finite-difference groundwater models using shared nodes: Advances in Water Resources, v. 25, no. 5, p. 497-511.

Mehl, S. and Hill, M.C., 2003, Locally refined block-centered finite-difference groundwater models: evaluation of parameter sensitivity and the consequences for inverse modeling: in, Kovar, K. and Hrkal, Z., IAHS Publication no. 277, p. 227-232.

Mehl, S. and Hill, M.C., 2004, Three-dimensional local grid refinement for block-centered finite-difference groundwater models using iteratively coupled shared nodes: a new method of interpolation and analysis of errors: Advances in Water Resources, v. 27, no 9, p. 899-912.

Nacul, E. C., 1991, Use of domain decomposition and local grid refinement in reservoir simulation, PhD thesis, Stanford University, 370 p.

Poeter, E. and Hill, M.C., 1998, Documentation of UCODE, A computer code for universal inverse modeling: U.S. Geological Survey Water-Resources Open-File Report 98-4080, 116 p.

Poeter, E., Hill, M.C., Banta, E. R., Mehl, S., and Christensen, S., 2005, UCODE_2005 and six other computer codes for universal sensitivity analysis, calibration, and uncertainty evaluation: U.S. Geological Survey Techniques and Methods 6-A11, 283 p. (Available at http://www.mines.edu/igwmc/freeware/ucode/)

Pollock, D.W., 1994, User's Guide for MODPATH/MODPATH-PLOT, Version 3: A particle tracking post-processing package for MODFLOW, the U.S. Geological

Survey finite-difference ground-water flow model: U.S. Geological Survey Open-File Report 94-464, 249 p. (Available at http://water.usgs.gov/nrp/gwsoftware/modpath41/modpath41.html)

Quandalle, P. and Besset, P., 1985, Reduction of Grid Effects Due to Local Sub-Gridding in Simulations Using a Composite Grid: Paper SPE 13527 presented at the Eighth SPE Symposium on Reservoir Simulation, Dallas, TX, Feb. 10-13, p. 295-305.

Schaars, F. and Kamps, P., 2001, MODGRID: Simultaneous solving of different groundwater flow models at various scales: In: Proceedings of MODFLOW 2001 and Other Modeling Odysseys Conference, Golden, CO, Sep. 11-14, vol I, p. 38-44.

Székely, F., 1998, Windowed Spatial Zooming in Finite-Difference Ground Water Flow Models: Ground Water, v. 36, no.5, p. 718-721.

von Rosenberg, D.U., 1982, Local Mesh Refinement for Finite Difference Methods: Paper SPE 10974 presented at the 57th Society of Petroleum Engineers Annual Fall Technical Conference, New Orleans, LA, Sep. 26-29.

Ward, D.S., Buss, D.R., Mercer, J.W., and Hughes, S.S., 1987, Evaluation of a Groundwater Corrective Action at the Chem-Dyne Hazardous Waste Site Using a Telescopic Mesh Refinement Modeling Approach: Water Resources Research, v. 23, no.4, p. 603-617.

Wasserman, M.L., 1987, Local Grid Refinement for Three-Dimensional Simulators, Paper SPE 16013 presented at the Ninth SPE Symposium on Reservoir Simulation, San Antonio, TX, Feb. 1-4, p. 231-241.

Weiser, A., and Wheeler, M.F., 1988, On convergence of block-centered finite differences for elliptic problems, SIAM Journal of Numerical Analysis, v. 23, no. 2, p. 351-375.

Wilson, J.D. and Naff, R.L., 2004, The U.S. Geological Survey modular ground-water model – GMG linear equation solver package documentation: U.S. Geological Survey Open-File Report 2004-1261, 47 p. (Available at http://water.usgs.gov/pubs/of/2004/1261/)

Zheng, C. and Wang, P., 1999, MT3DMS: A modular three-dimensional multi-species transport model for simulation of advection, dispersion and chemical reactions of contaminants in groundwater systems; Documentation and user's guide: Contract report SERDP-99-1: Vicksburg, Miss., U.S. Army Engineer Research and Development Center, 202 p. (Available at http://hydro.geo.ua.edu/mt3d)

Appendix 1 – LGR Input Instructions and Selected Input and Output Files from Examples 1 and 3

LGR Input Instructions

When executed, MODFLOW-2005 prompts for the name of a file. If a Name File (Harbaugh and others, 2000, p. 7, 43) is entered, LGR is not used. To use LGR, the name of the LGR Control File is entered. The contents of this file are described here.

The LGR Control File is distinguished from a Name file by the presence of a keyword "LGR" as the first non-commented input. LGR reads its input data from this file. Input for LGR is defined using 15 items. Each item is read free format.

FOR EACH SIMULATION

1. LGR

2. NGRIDS

FOR THE PARENT GRID (the parent grid needs to be listed before the child grid)

3. NAME FILE

4. GRIDSTATUS

5. IUPBHSV   IUPBFSV

FOR THE CHILD GRID

6. NAME FILE

7. GRIDSTATUS

8. ISHFLG   IBFLG   IUCBHSV   IUCBFSV

9. MXLGRITER   IOUTLGR

10.   RELAXH   RELAXF

11.   HCLOSELGR FCLOSELGR

12.   NPLBEG   NPRBEG   NPCBEG

13.   NPLEND   NPREND   NPCEND

14.   NCPP

15.   NCPPL (NPLEND +1 – NPLBEG)

Explanation of Variables Read by LGR

NGRIDS – is the number of grids used in this simulation. Currenlty (2005) NGRIDS needs to equal 2.

NAME FILE– is the name of the Name file for either the parent or child grid. The name can include the file path and is limited to 200 characters.

GRIDSTATUS – is a character variable indicating whether the file listed in NAME FILE corresponds to a parent or child grid.
    If GRIDSTATUS = PARENTONLY, then it is a parent grid Name file.
    If GRIDSTATUS = CHILDONLY, then it is a child grid Name file.

Appendix 1 – LGR Input Instruction and Selected Data Input and Output Files from Examples 1 and 3

IUPBHSV – a number greater than zero that corresponds to the unit number where the boundary heads are saved for later use by the BFH Package for independent simulations. A file with this unit number needs to be opened in the Name file of the parent model. A value of zero indicates that the file is not written. For the parent model, these are the complementary boundary conditions (see Appendix 2).

IUPBFSV – a number greater than zero that corresponds to the unit number where the boundary fluxes are saved for later use by the BFH Package for independent simulations. A file with this unit number needs to be opened in the Name file of the parent model. A value of zero indicates that the file is not written. For the parent model, these are the coupling boundary conditions (see Appendix 2).

ISHFLG – is a flag indicating whether heads from the parent grid simulation should be used as the starting head for the child grid simulation. These heads apply to the interior of the child, not the boundary.

> If ISHFLG = 1, then use results of the parent grid simulation as the starting head for the child grid. In the cells of the child grid that overlap the parent grid, the heads of the corresponding parent cell are used. No interpolation is applied. For steady-state simulations, this can provide a good initial guess which can reduce computational time. For transient simulations, this overwrites the initial condition of the child model defined in STRT of the Basic Package input file and therefore is not recommended.

> If ISHFLG = 0, then use the heads defined in STRT of the Basic Package for the child grid.

IBFLG – is a negative integer used to define the interface of the child grid with the parent. Use this value around the perimeter of the child model IBOUND array. Do not use IBFLG or -IBFLG anywhere else in the parent or child IBOUND arrays.

IUCBHSV – a number greater than zero that corresponds to the unit number where the boundary heads are saved for later use by the BFH Package for independent simulations. A file with this unit number needs to be opened in the Name file of the child model. A value of zero indicates that the file is not written. For the child model, these are the coupling boundary conditions (see Appendix 2).

IUCBFSV – a number greater than zero that corresponds to the unit number where the boundary fluxes are saved for later use by the BFH Package for independent simulations. A file with this unit number needs to be opened in the Name file of child model. A value of zero indicates that the file is not written. For the child model, these are the complementary boundary conditions (see Appendix 2).

MXLGRITER – is the maximum number of LGR iterations; 20 iterations are sufficient for most problems. See Closure Criteria for LGR Iterations section. Set MXLGRITER to 1 for a one-way coupling.

IOUTLGR – is a flag that controls printing from LGR iterations of the maximum head and flux change. For the maximum head change, the head value and corresponding layer, row, and column of the child grid is listed. For the maximum flux change, the flux value and corresponding layer, row, and column of the parent grid is listed. If IOUTLGR < 0, output is written to the screen. If IOUTLGR >0, output is written to the child listing file. If IOUTLGR = 0, no results are written.

RELAXH – is the relaxation factor for heads.

RELAXF – is the relaxation factor for fluxes.

Values of RELAXH and RELAXF less than 1 and greater than zero are needed for convergence of the LGR iterations. Typically, values around 0.5 produce convergent solutions. Values less than 0.5 may be needed when the LGR iterations have difficulty converging. In cases when the LGR iterations exhibit no convergence difficulties, values greater than 0.5 may reduce the number of iterations needed for convergence. Convergence difficulties can be diagnosed by printing the maximum head and flux changes (IOUTLGR $\neq$ 0) to determine if the head and flux changes are decreasing (converging) or increasing (diverging) as the LGR iterations proceed.

HCLOSELGR – is the head closure criterion for the LGR iterations. The closure criterion is based on heads of the child interface nodes. This closure criterion is satisfied when the maximum absolute head change between successive LGR iterations is less than HCLOSELGR (see equation 8b).

FCLOSELGR – is the flux closure criterion for the LGR iterations. The closure criterion is based on fluxes into the parent interface nodes. This closure criterion is satisfied when the maximum absolute relative flux change between successive LGR iterations is less than FCLOSELGR (see equation 8a).

NPLBEG – is the number of the topmost layer of the parent grid where the child model begins. Currently, (2005) refinement must begin at the top of the model so NPLBEG = 1.

NPRBEG – is the row number of the parent grid where the refinement begins (cannot equal 1).

NPCBEG – is the column number of the parent grid where the refinement begins (cannot equal 1).

NPLEND – is the number of the lowest layer of the parent grid where the refinement ends. NPLEND $\geq$ NPLBEG

NPREND – is the row number of the parent grid where the refinement ends. NPREND > NPRBEG and NPREND cannot equal the number of rows in the parent grid.

NPCEND – is the column number of the parent grid where the refinement ends. NPCEND > NPCBEG and NPCEND cannot equal the number of columns in the parent grid.

NCPP – the number of child cells that span the width of a single parent cell along rows and columns. This must be an odd integer number > 1 and is applied to rows and columns.

NCPPL – is the number of child cells that span the depth of a single parent layer. This must be an odd integer number $\geq$ 1. Read one value for each refined parent layer. The number of values needs to equals NPLEND +1 minus NPLBEG. Values can be 1, which results in no vertical refinement for the layer, only in layers above the bottom of the child grid, unless the refinement extends all the way to the bottom of the parent model (see The Top and Bottom of the Child Grid section). For refinement that does not extend to the bottom of the parent model, the refinement terminates at the shared node; for example, in Figure 5b the values 5 3 would be needed.

Example LGR Input Files

The sample data inputs listed below are for the two-dimensional example 1 and three-dimensional example 3 presented in the text of this report. The systems are shown in Figure 15 and Figure 23

Example 1 uses a 9:1 refinement ratio. The refinement begins in layer 1, row 20, column 22 and ends in layer 1, row 31, column 39 of the parent grid. The annotated LGR input file is:

```
LGR                     #LGR Keyword
2                       #NGRIDS
ex2_parent.nam          #NAME FILE
PARENTONLY              #GRIDSTATUS
70 71                   #Unit #'s for complimentary and coupling B.C.
ex2_child.nam           #NAME FILE
```

Appendix 1 – LGR Input Instruction and Selected Data Input and Output Files from Examples 1 and 3

CHILDONLY	#GRIDSTATUS
1 -59 80 81	#ISHFLG, IBFLG, Unit #'s for complimentary and coupling B.C.
15 0	#MXLGRITER, IOUTLGR
0.50 0.50	#RELAXH, RELAXF
1.0E-5 1.0E-5	#HCLOSELGR, FCLOSELGR
1 20 22	#Beginning layer, row, and column
1 31 39	#Ending layer, row, and column
9	#Horizontal refinement ratio
1	#Vertical refinement ratio by Parent layer

Example 3 is for a three-dimensional problem and uses a 3:1 ratio of refinement in all directions, in all layers. The refinement begins in layer 1, row 6, column 2 and ends in layer 2, row 12, column 9 of the parent grid (see Figure 23).

LGR	#LGR Keyword
2	#NGRIDS
ex3_parent.nam	#NAME FILE
PARENTONLY	#GRIDSTATUS
0 0	#Unit #'s for complimentary and coupling B.C.
ex3_child.nam	#NAME FILE
CHILDONLY	#GRIDSTATUS
1 -59 80 81	#ISHFLG, IBFLG, Unit #'s for complimentary and coupling B.C.
15 0	#MXLGRITER, IOUTLGR
0.40 0.40	#RELAXH, RELAXF
5.0E-3 5.0E-2	#HCLOSELGR, FCLOSELGR
1 6 2	#Beginning layer, row, and column
2 12 9	#Ending layer, row, and column
3	# Horizontal refinement ratio
3 3	# Vertical refinement ratio by Parent layer

Sample LGR Output

The output from MODFLOW-2005 when using LGR differs from standard MODFLOW-2005 output in two ways. First, the LGR control input file is echoed. Second, an overall volumetric budget includes interface fluxes. When LGR is active, the budgets for the parent and child grid simulations include the amount of flow into and out of the child grid through the interface of the parent grid (see Closure Criteria for LGR Iterations section). Using the three-dimensional example 3, the output appears as follows:

Parent model output:

Appendix 1 – LGR Input Instruction and Selected Data Input and Output Files from Examples 1 and 3

. . .

```
LGR -- LOCAL GRID REFINEMENT, VERSION 1.0, 02/15/2006
      INPUT READ FOR MODEL  1 DEFINED BY NAME FILE ex3_parent.nam

LOCAL GRID REFINEMENT IS ACTIVE FOR PARENT ONLY
```

. . .

```
VOLUMETRIC BUDGET FOR ENTIRE MODEL AT END OF TIME STEP  1 IN STRESS PERIOD   1
------------------------------------------------------------------------------

    CUMULATIVE VOLUMES      L**3          RATES FOR THIS TIME STEP      L**3/T
    ------------------      ----          ------------------------      ------

           IN:                                    IN:
           ---                                    ---
           STORAGE =          0.0000              STORAGE =          0.0000
     CONSTANT HEAD =        158.2277        CONSTANT HEAD =        158.2277
      RIVER LEAKAGE =       142.0706         RIVER LEAKAGE =       142.0706
   PARENT FLUX B.C. =        51.2186      PARENT FLUX B.C. =        51.2186

          TOTAL IN =        351.5168             TOTAL IN =        351.5168

           OUT:                                   OUT:
           ----                                   ----
           STORAGE =          0.0000              STORAGE =          0.0000
     CONSTANT HEAD =        154.0517        CONSTANT HEAD =        154.0517
      RIVER LEAKAGE =        22.3512         RIVER LEAKAGE =        22.3512
   PARENT FLUX B.C. =       175.5873      PARENT FLUX B.C. =       175.5873

         TOTAL OUT =        351.9903            TOTAL OUT =        351.9903

         IN - OUT =          -0.4734            IN - OUT =          -0.4734

PERCENT DISCREPANCY =          -0.13   PERCENT DISCREPANCY =          -0.13
```

Child model output:

. . .

```
LGR -- LOCAL GRID REFINEMENT, VERSION 1.0, 02/15/2006
      INPUT READ FOR MODEL  2 DEFINED BY NAME FILE ex3_child.nam

LOCAL GRID REFINEMENT IS ACTIVE FOR CHILD ONLY

              STARTING HEADS FROM PARENT WILL BE USED: ISHFLG =   1
              VALUE IN IBOUND INDICATING BOUNDARY INTERFACE = -59
              BOUNDARY HEADS WILL BE SAVED ON UNIT 80
              BOUNDARY FLUXES WILL BE SAVED ON UNIT 81
              MAX NUMBER OF LGR ITERATIONS = 15
              LGR ITERATIONS RESULTS NOT WRITTEN: IOUTLGR=  0

              WEIGHTING FACTORS FOR RELAXATION
              RELAXH(HEAD) RELAXF(FLUX)
              ------------------------
              0.400E+00       0.400E+00

              CLOSURE CRITERIA FOR LGR ITERATIONS
              HCLOSELGR       FCLOSELGR
              ------------------------
              5.000E-03       5.000E-02

              STARTING LAYER, ROW, COLUMN=   1,   6,   2
              ENDING LAYER, ROW, COLUMN=     2,  12,   9
              NCPP: NUMBER OF CHILD CELLS PER WIDTH OF PARENT CELL= 3
              NCPPL: NUMBER OF CHILD LAYERS IN LAYER  1 OF PARENT = 3
              NCPPL: NUMBER OF CHILD LAYERS IN LAYER  2 OF PARENT = 3
```

. . .

```
FLUX ACROSS PARENT-CHILD INTERFACE AT TIME STEP  1 IN STRESS PERIOD  1
-----------------------------------------------------------------------------

     CUMULATIVE VOLUMES      L**3      RATES FOR THIS TIME STEP      L**3/T
     ------------------                ------------------------

   TOTAL IN TO CHILD =      175.5873      TOTAL IN TO CHILD =       175.5873

 TOTAL OUT TO PARENT =       51.2186    TOTAL OUT TO PARENT =        51.2186
```

The interface budget and the constant-head budget differ in two ways. First, the constant-head budget includes all constant heads in the child model, while the interface budget only includes the constant heads used along the interface. Second, the constant-head budget does not include fluxes between constant head cells while the interface budget does.

Literature Cited

Harbaugh, A.W., Banta, E.R., Hill, M.C., and McDonald, M.G., 2000, MODFLOW-2000, the U.S. Geological Survey modular ground-water model – User guide to modularization concepts and the ground-water flow process: U.S. Geological Survey Open-File Report 00-92, 121 p.

Appendix 2 – Independent Simulations Using the Boundary Flow and Head (BFH) Package

The Boundary Flow and Head (BFH) Package reads input data from the file indicated in the Name file as described by Harbaugh and others (2000, p. 7, 43) using the File Type BFH. Input for the BFH Package is created by LGR and requires that the coupling boundary conditions calculated by LGR be saved using variable IUCBHSV and (or) IUPBFSV of the LGR input file. For an independent child model simulation, IUCBHSV needs to be nonzero; for an independent parent model simulation, IUPBFSV needs to be nonzero.

The BFH Package and LGR cannot be used simultaneously. Thus, when using LGR, the Name file specified in the LGR control file cannot use file type BFH.

The procedure needed to run independent child or parent models with LGR boundary conditions is as follows:

1) Use LGR to calculate and save the coupling boundary conditions.

2) Activate the BFH Package in the Name file with a file name that corresponds to the file saved on IUCBHSV or IUPBFSV for child and parent simulations, respectively.

As discussed in the Running the Parent and Child Model Independently Using the Boundary Flow and Head (BFH) Package section, the BFH package can be used to evaluate the effects of model changes on the boundary conditions. In this case, the complementary boundary conditions also need to be saved when running LGR. For the child model, IUCBFSV needs be nonzero; for the parent model, IUPBHSV needs to be nonzero. If the file containing the complementary boundary conditions for the child or parent models is opened in the Name file on the unit number corresponding to IUCBFSV or IUPBHSV, respectively, then the BFH package will evaluate the discrepancies in the complementary boundary conditions.

Each of these files contains a header record and a list of the child and parent cells involved in the coupling, indicated by the layer, row, and column. For the child models, the corresponding adjoining parent cells and a node index is listed with each child cell. This is followed by a listing of the boundary head or flux values, corresponding to these cells, for each time step.

Example BFH Inputs

The options for the BFH Package can be controlled through inputs to LGR and the Name files. Using the three-dimensional example 3 in Appendix 1, a simulation using LGR is performed first. For an independent simulation of the child grid, the coupling boundary condition (specified head) is saved on unit 80 and the complementary boundary condition (boundary flux) is saved on unit 81.

The Name file for the child grid for the LGR simulation is:

```
LIST     26 ex3_child.out
BAS6      2 ex3_child.ba6
BCF6     21 ex3_child.bc6
DIS      29 ex3_child.dis
OC       20 ex3_child.oc
DATA(BINARY) 31 ex3_child.hed
DATA(BINARY) 41 ex3_child.flw
PCG      22 ex3_child_3.pcg
RIV      25 ex3_child.riv
DATA 80 ex3_child_bfh.hed
DATA 81 ex3_child_bfh.flw
DATA 51 ex3_child.bot
```

After successful completion of an LGR simulation, the child model can be simulated independently using the BFH Package. Only the Name file of the child grid needs to be modified. Activate BFH with a file name corresponding to the file where the coupling boundary conditions were saved. Although not required, in the example above, the complimentary boundary conditions were saved. If this file is opened in the Name file on the same unit number on which it was saved, the BFH Package will report any changes in the boundary fluxes of the child model. This is done in the example below. Use of # in the first column results in the line being ignored.

```
LIST     26 ex3_child.out
BAS6      2 ex3_child.ba6
BCF6     21 ex3_child.bc6
DIS      29 ex3_child.dis
OC       20 ex3_child.oc
DATA(BINARY) 31 ex3_child.hed
DATA(BINARY) 41 ex3_child.flw
PCG      22 ex3_child_3.pcg
RIV      25 ex3_child.riv
BFH 80 ex3_child_bfh.hed
#DATA 80 ex3_child_bfh.hed
DATA 81 ex3_child_bfh.flw
DATA 51 ex3_child.bot
```

Sample BFH Output

The output from MODFLOW-2005 when using the BFH Package will show the volumetric budget contributions from the coupling boundary conditions. If the complementary boundary conditions are saved by LGR and activated in the Name file, then the BFH Package reports changes in the complementary boundary conditions. That is, for the child model, which is coupled using specified-head boundary conditions, changes in flow through the interfacing boundary are reported. For the parent model, which is coupled using specified-flux boundary conditions, changes in head along the interfacing boundary are reported. Using the three-dimensional example above, the additions to the MODFLOW-2005 output for an independent child grid simulation appears as:

. . .

```
BFH -- BOUNDARY FLOW AND HEAD PACKAGE, VERSION 1.0, 02/15/2006
        INPUT READ FROM UNIT   80
  CHILD HEAD B.C.
RUNNING CHILD MODEL WITH   730 SPECIFIED HEAD BOUNDARY NODES
  CHECKING AGAINST FLUX BOUNDARY CONDITIONS ON UNIT   81
```

. . .

```
VOLUMETRIC BUDGET FOR BFH SPECIFIED HEADS AT TIME STEP  1 IN STRESS PERIOD   1
--------------------------------------------------------------------------------

       CUMULATIVE VOLUMES      L**3        RATES FOR THIS TIME STEP    L**3/T
       ------------------                  ------------------------


            TOTAL IN =     176.0170            TOTAL IN =       176.0170


            TOTAL OUT =     50.9548            TOTAL OUT =        50.9548



  BFH: BOUNDARY FLUX COMPARISON
  -----------------------------
  NEW TOTAL BOUNDARY FLUX =   125.062187
  OLD TOTAL BOUNDARY FLUX =   124.368706
  AVERAGE ABSOLUTE FLUX DIFFERENCE =  0.991417468E-02
  MAXIMUM ABSOLUTE FLUX DIFFERENCE OF  0.729522705E-01
  OCCURS AT PARENT LAYER 2 ROW 9 COLUMN 5
  NEW FLUX AT THIS NODE =  -9.00437832
  OLD FLUX AT THIS NODE =  -9.07733059
```

There are some small discrepancies in the boundary fluxes even though no modifications were made to the child model. However, these errors are on the same order of magnitude as the LGR closure criterion for fluxes (FCLOSELGR), which was 5.000E-02 for this simulation (Appendix 1). Furthermore, the volumetric budget error for the parent model, which indicates the overall quality of the LGR solution, is 0.13 percent (Appendix 1). Therefore, the small discrepancies reported by the BFH Package are expected. As the LGR closure criteria are decreased, the discrepancies decrease. Using the BFH Package in this way provides an indicator of the quality of the LGR solution.

If the child model is modified, the BFH Package can be used to assess the effects on the coupling boundary conditions. For example, consider changing the child model to include pumping at a rate of 9.0 m^3/day from layer 2, row 8, column 12 and simulated with the BFH Package. The results are shown below:

. . .

```
VOLUMETRIC BUDGET FOR BFH SPECIFIED HEADS AT TIME STEP   1 IN STRESS PERIOD    1
--------------------------------------------------------------------------------

       CUMULATIVE VOLUMES      L**3        RATES FOR THIS TIME STEP      L**3/T
       ------------------                  ------------------------

          TOTAL IN =      180.9922               TOTAL IN =        180.9922

          TOTAL OUT =      50.3717               TOTAL OUT =        50.3717


   BFH: BOUNDARY FLUX COMPARISON
   -----------------------------
   NEW TOTAL BOUNDARY FLUX =   130.620560
   OLD TOTAL BOUNDARY FLUX =   124.368706
   AVERAGE ABSOLUTE FLUX DIFFERENCE =  0.776892602E-01
   MAXIMUM ABSOLUTE FLUX DIFFERENCE OF   4.10331678
   OCCURS AT PARENT LAYER 2 ROW 8 COLUMN 6
   NEW FLUX AT THIS NODE =   11.9733610
   OLD FLUX AT THIS NODE =   7.87004423
```

Although the pumping well is located below a river node, only about 3 m^3/day comes from the river; the remaining 6 m^3/day come from the boundaries. Of this, about 4 m^3/day comes from the cells adjoining the parent cell in layer 2, row 8, column 6. The child cells that correspond to this parent cell can be determined from the complementary boundary condition file where the child and corresponding parent cells are listed.

Literature Cited

Harbaugh, A.W., Banta, E.R., Hill, M.C., and McDonald, M.G., 2000, MODFLOW-2000, the U.S. Geological Survey modular ground-water model – User guide to modularization concepts and the ground-water flow process: U.S. Geological Survey Open-File Report 00-92, 121 p.

Appendix 3 – Error Propagation in LGR

For the case where specified-head boundary conditions are used around the perimeter of the child grid, as in LGR, and when the governing ground-water flow equation is linear, we show that the error in the specified heads propagate into the interior by the governing equation for the aquifer system being modeled, minus the sink and source terms. First, define the specified-head boundary condition as the true head plus some error (from grid resolution, interpolation, and so on):

$$\{h_b\} = \{h_{Tb}\} + \{e\} \tag{12}$$

where,

$\{h_b\}$ = head boundary condition,

$\{h_{Tb}\}$ = true head at the boundary, and

$\{e\}$ = error at the boundary

The matrix equations resulting from a finite-difference discretization can be written as:

$$[A]\{h\} = \{C \cdot h_b + q\} \tag{13}$$

where,

> $[A]$ = the standard coefficient matrix resulting from a finite-difference discretization,
>
> $\{h\}$ = head in the child grid,
>
> C = a coefficient multiplying h_b which accounts for the conductance between the specified-head boundary condition and the aquifer, and
>
> $\{q\}$ = all other sink and source terms in the child grid

Substituting equation 12 into the right-hand side of equation 13:

$$[A]\{h\} = \{C \cdot (h_{Tb}+e) + q\} \tag{14}$$

Reordering the right-hand side, the solution can be written as:

$$\{h\} = [A]^{-1}\{C \cdot h_{Tb}+q+C \cdot e\} \tag{15}$$

The matrix multiplication can be distributed across the terms of the right-hand side:

$$\{h\} = [A]^{-1}\{C \cdot h_{Tb}+q\} + [A]^{-1}\{C \cdot e\} \tag{16}$$

which can be written as,

$$\{h\} = \{h_T\} + \{h_e\} \tag{17}$$

where,

> $\{h_T\} = [A]^{-1}\{C \cdot h_{Tb}+q\}$, and
>
> $\{h_e\} = [A]^{-1}\{C \cdot e\}$

This result stems directly from the principle of superposition where the effects of two components are added together (see Reilly and others, 1987). The first term on the right-hand side of equation 17 is the true head solution, $\{h_T\}$, which would be obtained if the true boundary conditions were used, $\{h_{Tb}\}$. The second term on the right-hand side, $\{h_e\}$, has the same coefficient matrix and represents how the additional error on the boundary is diffused through the grid. Because there are no sinks or sources for this second term – the sinks/sources are accounted for in the first term – it contains the boundary errors only. This results in two important properties: (1) the maximum error occurs at the boundary and (2) the error is propagated from the boundary through the grid by a purely diffusive process with the same coefficients of the ground-water flow system. The diffusion process causes a smoothing effect by averaging with neighboring cells. Thus, positive and negative errors on the boundary tend to cancel as they propagate into the interior. If the error is constant along the boundary, there will be no cancellation by averaging with neighboring cells, and the error is propagated directly into the interior.

The above analysis is strictly valid only when the governing ground-water flow equation is linear and the principle of superposition holds. For nonlinear situations, such as flow in unconfined aquifers, the errors are still propagated through the grid by a diffusion process; however the coefficients in the matrix are not identical to those from the ground-water flow system.

The above analysis can be used to evaluate the errors from transient simulations because the numerical solution to the transient ground-water flow equations can be viewed as a series of steady-state solutions. The error after the first time step is the same as that outlined above except that the coefficient matrix $[A]$ and right-hand side include a storage term. The error at subsequent time steps include diffusion of errors in the interior from the previous time step, plus the error introduced at the boundary at the current time step. Starting from equation 13, the modification for transient flow is:

$$[A]\{h^n\} = \{C \cdot h_b + S \cdot h^{n-1} + q^n\} \qquad (18)$$

where,

superscripts ^{n} and $^{n-1}$ denote the current time step and previous time step, respectively, and

S = a coefficient multiplying the heads at the previous time which accounts for the changes in storage during the time step.

Substituting equation 12 for h_b and equation 17 for h^{n-1} in equation 18 and following the previous derivation:

$$\{h^n\} = [A]^{-1}\{C \cdot h_{Tb} + S \cdot h_T^{n-1} + q^n\} + [A]^{-1}\{C \cdot e^n + S \cdot h_e^{n-1}\} \qquad (19)$$

The first term on the right-hand side of equation 19 is the true head solution that would be obtained if true boundary conditions were used in this time step, and the previous solution also were true. The second term represents the error due to errors at the boundary of this time step plus the errors that were propagated into the interior from the previous time step. Setting the storage term, S, to zero in equation 19 simplifies to the steady-state solution shown in equation 16, indicating that the errors will approach those given in equation 16 as the solution approaches steady state.

Literature Cited

Reilly, T.E., Franke, O.L., and Bennett, G.D., 1987, The principle of superposition and its application in ground-water hydraulics: U.S. Geological Survey Techniques of Water-Resources Investigations, book 3, chap. B6, 28 p.

Appendix 4 – Using MODTMR and MF96TO2K to Create Child-Model Input Files

Users can use MODTMR (Leake and Claar, 1999) to construct child-model data sets based on a parent model by specifying the grid coordinates and refinement such that there is an odd ratio of refinement and shared nodes, as required by LGR. The MF96TO2K conversion utility distributed with MODFLOW-2000 can be used to further modify the data sets generated by MODTMR to be compatible with MODFLOW-2005. Thicknesses are not explicitly defined in MODFLOW-96; therefore users should look carefully how thickness-dependent properties of the BCF Package, such as transmissivity, are converted and make necessary adjustments.

Literature Cited

Leake, S.A. and Claar, D.V., 1999, Procedure and computer programs for telescopic mesh refinement using MODFLOW: U.S. Geological Survey Open-File Report 99-238, 53 p.

Appendix 5 – Brief Program Description

Variables in Fortran Module LGRMODULE

Table 5-1. Variables in Fortran module LGRMODULE

Variable Name	Size	Description
ISCHILD	Scalar	Flag: -1=parent grid, 1= child grid.
NPLBEG	Scalar	Layer in the parent grid where refinement begins.
NPRBEG	Scalar	Row in the parent grid where refinement begins.
NPCBEG	Scalar	Column in the parent grid where refinement begins.
NPLEND	Scalar	Layer in the parent grid where refinement ends.
NPREND	Scalar	Row in the parent grid where refinement ends.
NPCEND	Scalar	Column in the parent grid where refinement ends.
NCPP	Scalar	Refinement ratio along rows and columns.
NPL	Scalar	Number of parent layers that are refined.
IBOTFLG	Scalar	Flag: 0=bottom layer is not refined, 1=bottom layer is refined.
ISHFLG	Scalar	Flag: 0=do not use initial parent solution for interior of child, 1=use initial parent solution for interior of child.
IBLFG	Scalar	Unit number used in child IBOUND array that denotes the interface boundary.
IUPBHSV	Scalar	Unit number where parent boundary heads are saved for use with the BFH Package.
IUCBHSV	Scalar	Unit number where child boundary heads are saved for use with the BFH Package.
IUPBFSV	Scalar	Unit number where parent boundary fluxes are saved for use with the BFH Package.
IUCBFSV	Scalar	Unit number where child boundary fluxes are saved for use with the BFH Package.
MXLGRITER	Scalar	Maximum number of LGR iterations allowed.
IOUTLGR	Scalar	Flag: -1=print LGR iterations to screen, 1=print LGR iterations to listing file, 0=do not print LGR iterations.
RELAXH	Scalar	Relaxation factor for heads.
RELAXF	Scalar	Relaxation factor for fluxes.
HCLOSELGR	Scalar	Head closure criterion for LGR iterations.
FCLOSELGR	Scalar	Flux closure criterion for LGR iterations.
HDIFFM	Scalar	Maximum absolute head change between successive LGR iterations.
FDIFFM	Scalar	Maximum absolute flux change between successive LGR iterations.
NCPPL	NPL	Vertical refinement ratio for each parent layer that is refined.
NODEH	3	NODEH(n) identifies cell with maximum head change (HDIFFM): n=1 – Layer number. n=2 – Row number. n=3 – Column number.
NODEF	3	NODEF(n) identifies cell with maximum flux change (FDIFFM): n=1 – Layer number. n=2 – Row number. n=3 – Column number.
KPLC	Number of child boundary nodes	Array that maps the index of the child interface cell to the corresponding parent layer number. The index is defined by looping through the child interface cells in the order of columns, rows, and layers.
IPLC	Number of child boundary nodes	Array that maps the index of the child interface cell to the corresponding parent row number. The index is defined by looping through the child interface cells in the order of columns, rows, and layers.

JPLC	Number of child boundary nodes	Array that maps the index of the child interface cell to the corresponding parent column number. The index is defined by looping through the child interface cells in the order of columns, rows, and layers.
NPINDX	Number of boundary nodes	Array that maps an index of child interface cells to an index of the parent interface cells based on looping through the interface cells in the order of columns, rows, and layers.
ICBOUND	NODES (child model)	A copy of the child IBOUND array.
HOLDC	NCOL,NROW,NLAY (child model)	Head in the child grid at the previous LGR iteration.
CCC	NCOL,NROW,NLAY (child model)	Copy of the child conductance along columns.
CCR	NCOL,NROW,NLAY (child model)	Copy of the child conductance along rows.
CCV	NCOL,NROW,NLAY (child model)	Copy of the child vertical conductance.
PFLUX	(NPCBEG:NPCEND, NPRBEG:NPREND, NPLBEG:NPLEND)	Flux across the parent-child interface, accumulated to parent-grid cells.
PFLUXOLD	(NPCBEG:NPCEND, NPRBEG:NPREND, NPLBEG:NPLEND)	Flux across the parent-child interface at the previous LGR iteration. Accumulated to parent-grid cells.
VCB	4	Volumetric budget values for the interface specified-head cells of the child grid: (1) – Inflow rate for current time step. (2) – Outflow rate for current time step. (3) – Cumulative volume of inflow. (4) – Cumulative volume of outflow.

Description of LGR Subroutines

Listed below are the subroutines that are within the gwf2lgr1.f file. Subroutines that are called from main are in bold, underlined text.

GETNAMFILLGR – Reads the Name files from the LGR control file.

GWF2LGR1AR – Allocates and reads data for LGR. Calls SGWF2LGR1PSV to save pointer arrays for LGR data.

GWF2LGR1DA – Deallocates LGR module data.

GWF2LGR1RP – Finds the mapping between the column, row, and layer of interface child cell to the corresponding location of the parent grid. Calls SGWF2LGR1PNT to change pointers for LGR data to the appropriate grid.

GWF2LGR1INITP – Zeros out the interior cells of the parent grid which are completely covered by the child cells. Called after the first full parent solution.

GWF2LGR1FMCBS – Adjusts the child interface boundary storage coefficients. Calls SGWF2LGR1FMCS which does the adjustment based on which flow package is active by calling SGWF2LGR1BCFSA if BCF is active, SGWF2LGR1LPFSA if LPF is active, or SGWF2LGR1HUFSA if HUF is active.

GWF2LGR1FMPBS – Adjusts the parent interface boundary storage coefficients. Calls SGWF2LGR1FMPS which does the adjustment based on which flow package is active by calling SGWF2LGR1BCFSA if BCF is active, SGWF2LGR1LPFSA if LPF is active, or SGWF2LGR1HUFSA if HUF is active.

GWF2LGR1FMIB – Adjusts the child interface boundary IBOUND, and the child interior CC, CR, and CV arrays for the cage-shell interpolation. Calls SGWF2IBSHARE1 to set up the cage solution and SGWF2IBSHARE2 to set up the shell solution.

GWF2LGR1FMCBC – Adjusts the child interface boundary conductances.

GWF2LGR1FMPBC – Adjusts the parent interface boundary conductances.

GWF2LGR1BH – Transfers the parent interface boundary heads to the shared nodes of the child grid. Calls SLGR1HRLX to relax the head change at the shared nodes and find the location and value of the maximum head change.

GWF2LGR1FMBF – Calculates child interface boundary fluxes and creates the specified flux boundaries for the parent grid. Depending on which flow package is active, calls SGWF2LGR1BCFSF, or SGWF2LGR1LPFSF, or SGWF2LGR1HUFSF to find flux contribution from changes in storage. Calls SLGR1FRLX to relax the flux change and find the location and value of the maximum flux change.

GWF2LGR1CNVG – Checks for convergence of the LGR iterations. Calls SGWF2LGR1PNT to change pointers for LGR data to the appropriate grid.

GWF2LGR1PBD – Calculates the volumetric budget of the parent interface boundary flux. Calls SGWF2LGR1BFHPOT to output parent interface boundary heads and fluxes for use with the BFH Package.

GWF2LGR1CBD – Calculates the volumetric budget of the child interface boundary specified heads. Calls SGWF2LGR1BFHCOT to output child interface boundary heads and fluxes for use with the BFH Package

Variables in Fortran Module GWFBFHMODULE

Table 5-2. Variable in Fortran module GWFBFHMODULE.

Variable Name	Size	Description
ISCHILD	Scalar	Flag: -1=parent grid, 1= child grid.
NPLBEG	Scalar	Layer in the parent grid where refinement begins.
NPRBEG	Scalar	Row in the parent grid where refinement begins.
NPCBEG	Scalar	Column in the parent grid where refinement begins.
NPLEND	Scalar	Layer in the parent grid where refinement ends.
NPREND	Scalar	Row in the parent grid where refinement ends.
NPCEND	Scalar	Column in the parent grid where refinement ends.
NCPP	Scalar	Refinement ratio along rows and columns.
IBOTFLG	Scalar	Flag: 0=bottom layer is not refined, 1=bottom layer is refined.
IBLFG	Scalar	Unit number used in child IBOUND array that denotes the interface boundary.
NBNODES	Scalar	Number of interface boundary nodes in the child grid.
NPBNODES	Scalar	Number of interface boundary nodes in the parent grid.
NTIMES	Scalar	Number of times the interface boundary conditions are saved.
IUBC	Scalar	Flag and unit number. If IUBC ≠ 0, it is the unit number of the file where the complimentary boundary conditions are saved.
BTEXT	C*17	Name of the boundary condition.
KLAY	NBNODES	Array that maps an index of child interface boundary cell to the corresponding layer number of the child. The index is defined by looping over the interface cells in the order of columns, rows, and layers.
IROW	NBNODES	Array that maps an index of child interface boundary cell to the corresponding row number of the child. The index is defined by looping over the interface cells in the order of columns, rows, and layers.

JCOL	NBNODES	Array that maps an index of child interface boundary cell to the corresponding column number of the child. The index is defined by looping over the interface cells in the order of columns, rows, and layers.
KPLAY	NBNODES	Array that maps the index of the child interface cell to the corresponding parent layer number. The index is defined by looping through the child interface cells in the order of columns, rows, and layers.
IPROW	NBNODES	Array that maps the index of the child interface cell to the corresponding parent row number. The index is defined by looping through the child interface cells in the order of columns, rows, and layers.
JPCOL	NBNODES	Array that maps the index of the child interface cell to the corresponding parent column number. The index is defined by looping through the child interface cells in the order of columns, rows, and layers.
NPINDX	NBNODES	Array that maps an index of child interface cells to an index of the parent interface cells based on looping through the interface cells in the order of columns, rows, and layers.
BFLUX	NPBNODES	Flux across the parent-child interface. Used as the coupling boundary condition for the parent grid.
BFLUXCHK	NPBNODES	Flux across the parent-child interface. Used as the complimentary boundary condition for the child grid.
BHEAD	NBNODES	Head at the child interface boundary. Used as the coupling boundary condition for the child grid.
BHEADCHK	NBNODES	Head at the child interface boundary. Used as the complimentary boundary condition for the parent grid.
VCB	4	Volumetric budget values for the interface specified heads of the child grid: (1) – Inflow rate for current time step. (2) – Outflow rate for current time step. (3) – Cumulative volume of inflow. (4) – Cumulative volume of outflow.

Description of BFH Subroutines

Listed below are the subroutines that are within the gwf2bfh1.f file. Subroutines that are called from main are in bold, underlined text.

GWF2BFH1AR – Allocates and reads data for the BFH Package. Calls SGWF2BFH1PSV to save pointer arrays for BFH data.

GWF2BFH1DA – Deallocates BFH module data.

GWF2BFH1RP – Reads the column, row, and layer indices for cells along the interface where the boundary conditions are applied. Calls SGWF2LGR1PNT to change pointers for BFH data to the appropriate grid. Calls SGWF2BFH1FMCS to adjust child grid storage coefficients or SGWF2BFH1FMPS to adjust parent grid storage coefficients. These routines call SGWF2BFH1BCFSA if BCF is active, SGWF2BFH1LPFSA if LPF is active, or SGWF2BHF1HUFSA if HUF is active to adjust the storage. Zeros out the interior cells of the parent grid that will be completely covered by the child grid.

GWF2BFH1AD – Reads the coupling boundary condition data for the current time step and also the complimentary boundary condition data, if used. If it is a child grid, apply the specified head boundary conditions to HNEW.

GWF2BFH1FM – Calls SGWF2BFH1FMCBC or SGWF2BFH1FMPBC to adjust conductances along the parent-child boundary interface for the child or parent grid, respectively. For the parent grid, calls SGWF2BFH1FMPBF apply the parent interface boundary specified flux to RHS.

Appendix 5 – Brief Program Description

GWF2BFH1BD – Calculates the volumetric budget for the parent interface boundary specified fluxes and the child interface boundary specified heads. For the parent grid, the fluxes are added to the global budget accumulators. Calls SGWF2BFH1PNT to change pointers for BFH data to the appropriate grid. For the child grid, calls SGWF2BFH1CBF to find budgets for interface boundary specified heads. Call SGWF2BFH1BCFSF if BCF is active, SGWF2BFH1LPFSF if LPF is active, or SGWF2BFH1HUFSF if HUF is active to find the flux contribution from changes in storage.

GWF2BFH1OT – this subroutine calls SGWF2BFH1CBD to output to the listing file a separate budget for the child interface boundary specified heads when global budgets are printed. If complimentary boundary conditions are used for either the parent or child grid, then report the location and value of the maximum difference.

www.ingramcontent.com/pod-product-compliance
Lightning Source LLC
Chambersburg PA
CBHW081601170526
45166CB00009B/2775